The Metaparadigm of Clinical Dietetics:

Derivation and Applications

The Metaparadigm of Clinical Dietetics: Derivation and Applications

Ruth Leyse-Wallace PhD, RD

iUniverse, Inc.
New York Lincoln Shanghai

The Metaparadigm of Clinical Dietetics: Derivation and Applications

iUniverse books may be ordered through booksellers or by contacting:

iUniverse
2021 Pine Lake Road, Suite 100
Lincoln, NE 68512
www.iuniverse.com
1-800-Authors (1-800-288-4677)

The information, ideas, and suggestions in this book are not intended as a substitute for professional medical advice. Before following any suggestions contained in this book, you should consult your personal physician. Neither the author nor the publisher shall be liable or responsible for any loss or damage allegedly arising as a consequence of your use or application of any information or suggestions in this book.

ISBN-13: 978-0-595-42205-0 (pbk)
ISBN-13: 978-0-595-86545-1 (ebk)
ISBN-10: 0-595-42205-5 (pbk)
ISBN-10: 0-595-86545-3 (ebk)

Printed in the United States of America

Author's permission is hereby granted to use the ideas and materials in this publication for professional research and educational purposes. Permission is granted for nonprofit use only.

Please extend the professional courtesy of notifying me of the uses of the material and ideas found herein. Thank you.

_____

Ruth Leyse-Wallace, PhD, RD
email: RthLys@cox.net

Date_____

Dear Dr. Leyse-Wallace,

I plan to use ideas or materials concerning The Metaparadigm of Clinical Dietetics/Metaparadigm of Dietetics in the following manner:

_____

_____

_____

_____

_____

Sincerely,

Name _____

Address_____

_____

_____

_____

The original dissertation: <u>Perceptions of the Metaparadigm of Clinical Dietetics: Conceptual Delineation of Phenomena Relevant to the Discipline</u> by Ruth L. Leyse, 1998, may be ordered from UMI Dissertation Services, 300 North Zeeb Road, P. O. Box 1346, Ann Arbor Michigan. 1998. 1-800-521-0600. http://www.umi.com

Copies were deposited in the main library of the University of Arizona as well as the Department of Nutritional Sciences at the University of Arizona, Tucson, Arizona.

Contents

Section B—Applications of the Metaparadigm of Clinical Dietetics

Appendices

Acknowledgements

Many thanks and credit goes to Mary Ann Kight PhD, NIES. Her Research Typology for Clinical Dietetics was the cornerstone of the Metaparadigm of Clinical Dietetics. Dr. Kight's ideas, standards, enthusiasm and support as mentor and graduate school advisor are fully appreciated.

Thanks also goes to Ronna Biesecker, PhD, RD, a peer, fellow graduate student and colleague, for mutual support during the time we both entered graduate school at The University of Arizona after being immersed in the world of clinical practice and patient care for many years.

Appreciation also goes to the expert panel of clinical dietitians in Tucson, Arizona who shared their minds, spirit and time shaping the original research work over several years: Ronna Beisecker, PhD, RD; Dorothy Close, MS, RD; Marianna Gammon, MS, RD; Adele Huls, PhD, RD; Diane Parrington, PhD, RD; Mary Picchioni, MS, RD; and Annette Zagaroli, MS, RD.

In addition, appreciation goes to the clinical dietitians who participated in interviews and who tested the survey instrument and provided comments. Thanks also goes to the dietitians who participated as respondents to the survey and provided the data with which the Metaparadigm of Clinical Dietetics was validated.

Special thanks goes to Pamela Reed, PhD, FACN, RN, and Carrie Jo Braden, PhD, FACN, RN in the School of Nursing, intellectually stimulating professors and educators, as well as other members of my graduate committee Ralph L. Price, PhD, Lawrence M. Aleamoni, PhD, and Beth Stewart, PhD, RD.

Appreciation also goes to the principals of the Fairchild-Kight Endowment Fund for financial support of the dissertation described in this work.

Preface

This book has two main purposes. The first is to present in sufficient detail the development and testing of The Metaparadigm of Clinical Dietetics so that others can understand the scientific grounding of this work and extend its validation to additional groups or circumstances. The Metaparadigm of Clinical Dietetics can be used as defined by the original research, or may grow and develop as changes occur in the profession, health care, education and science. Validation with additional sub-sets of practitioners would lend it strength.

Defining professional knowledge in terms of a metaparadigm structure will make it more visible, easier to discuss, conceptualize, develop and communicate. As a profession, we have not been accustomed to having an organizational structure for our body of knowledge. The discipline has borrowed from, and shared knowledge with, the fields of nutrition science, education, nursing, medicine, public health, psychology, pharmacy and exercise physiology, to mention a few. Viewing knowledge in an organized manner has the potential for promotion of new profession-specific knowledge. Research developing the Metaparadigm of Clinical Dietetics indicates what clinical dietetic practitioners perceive to be the unique aspects of clinical dietetics. Findings in this work have the potential for contributing toward further evolution in the differentiation of clinical dietetics, other health professionals and nutritional scientists.

The second purpose is to illustrate how the Metaparadigm of Clinical Dietetics may be used by clinical dietetic practitioners, educators and researchers. The hypothetical examples presented may be used for theoretical and practical thinking in terms of domains of professional concern, for structuring a clinical practice, or for analyzing and planning professional educational needs—either for an individual or an institution.

It is possible The Metaparadigm of Clinical Dietetics can be expanded to the Metaparadigm of Dietetics. Mapping topics of articles published in The Journal of the American Dietetic Association, 2001, onto the domains of Metaparadigm of Clinical Dietetics indicates the appropriate fit of such domains to other areas of dietetics. An example of this mapping is included for future development.

In addition, there is an additional potential for this work: the development of the links between the Metaparadigm of Clinical Dietetics and the Eight World Hypotheses of Gary Schwartz, PhD and Linda Russek, PhD. This work contains just a glimmer of how the world view(s) of clinical dietetics link to these fascinating theories concerning how science has grown more inclusive and complex in its description of the world and how it works.

The material in this book is not an end point; it merely points a way toward clinical dietetic knowledge of the future.

Ruth Leyse-Wallace, PhD, RD

Introduction

Defining Knowledge

What is a Metaparadigm?

A metaparadigm is a concept naming one of four levels of knowledge. A metaparadigm is the most abstract of these four levels (Leyse and Kight, 1993). Metaparadigms encompass the concepts included in paradigms of the discipline. Paradigms are expressed through theories. Theories are statements of how the world is thought to be organized, stated in observable or measurable terms. Even though science is often thought of as the total process of discovering knowledge and describing the world, science for this purpose is defined as the most concrete level, the measurable level, of knowledge. Concrete knowledge is the measured or observed fact(s).

Examples of Continuum from Abstract to Concrete			
Abstract ⟵			⟶ **Concrete**
Human.	Girl	Ruth	Fingerprints, voice pattern, iris pattern
Light	Color.	Green.	Wavelength frequency
Furniture	Chair.	Booster Chair. . .	Brand and Model number
Metaparadigm . .	Paradigms . .	Theories.	Measurable Data, Observations

These four levels of knowledge can also be described as a hierarchy or as a continuum of knowledge. This work is a metatheory, or statement of how the knowledge utilized by the discipline of Clinical Dietetics can be organized.

Unorganized knowledge can be thought of as being in chaos. For order to come from chaos, a form or structure emerges, often with the help of a theorist who looks to see patterns of organization in the seeming chaos of life. (Wheatley, 1999) The structure assists scientists, practitioners and others to use, organize, link and build upon the knowledge. (Rubenstein, 1984) The practice of clinical dietetics uses a large quantity of knowledge. Dietetic practitioners have discovered some knowledge, some is unique to clinical dietetics, much is borrowed or shared with nutritional science and/or other health professionals. Some clinical dietitians perceive the role of the profession as the application of the knowledge discovered by nutritional scientists to the lives of individual clients. Others believe that clinical dietetic practitioners can expand the profession by enlarging the body of knowledge unique to the profession, by investigating observations made in practice, by creating and scientifically investigating theories found in the laboratory of clinical practice and nutritional patient care.

The Metaparadigm of Clinical Dietetics is the most abstract statement of the organizational structure of the profession's body of knowledge. (Leyse, 1998) A graphic depiction from most abstract to most concrete of this structure is shown below:

<u>The Body of Knowledge of Clinical Dietetics</u>

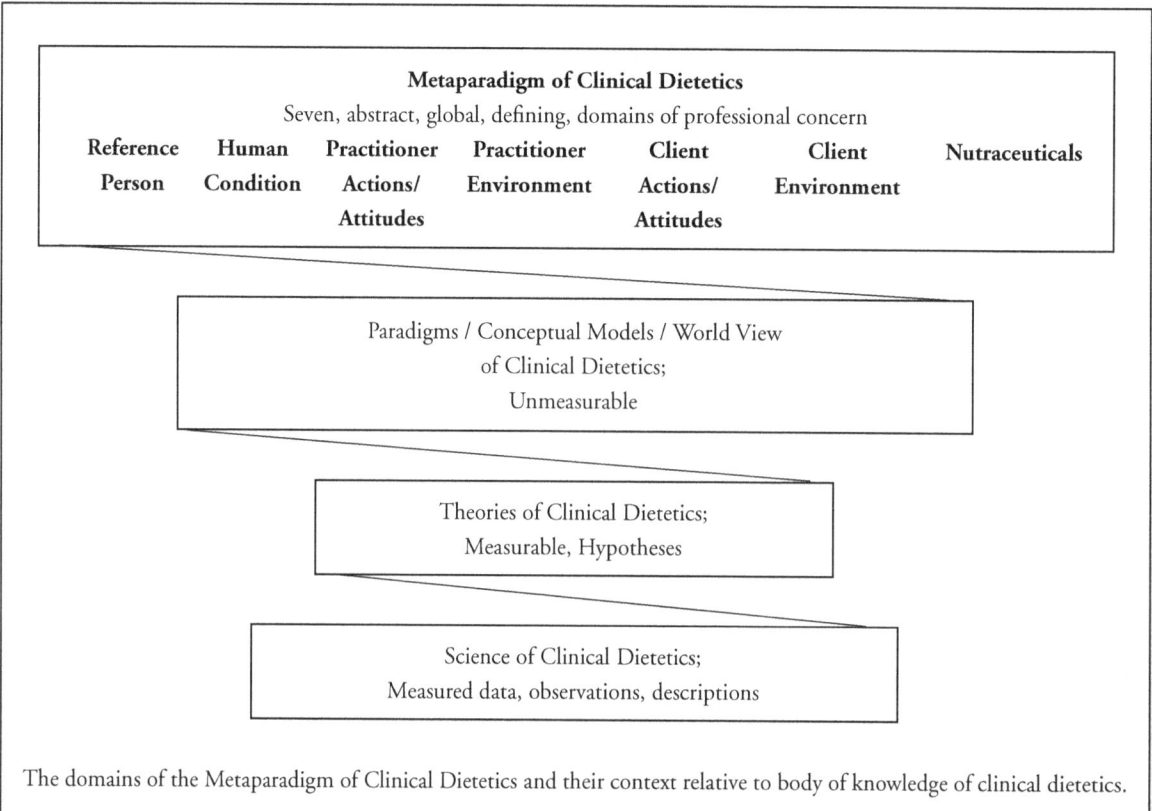

Metaparadigm of Clinical Dietetics
Seven, abstract, global, defining, domains of professional concern

| Reference Person | Human Condition | Practitioner Actions/ Attitudes | Practitioner Environment | Client Actions/ Attitudes | Client Environment | Nutraceuticals |

Paradigms / Conceptual Models / World View
of Clinical Dietetics;
Unmeasurable

Theories of Clinical Dietetics;
Measurable, Hypotheses

Science of Clinical Dietetics;
Measured data, observations, descriptions

The domains of the Metaparadigm of Clinical Dietetics and their context relative to body of knowledge of clinical dietetics.

The concepts of the Metaparadigm of Clinical Dietetics can be said to be the expression of the global concerns of Clinical Dietetics. These concerns can be thought of as the phenomena of concern to the profession. These concepts state what Clinical Dietetic practitioners concern themselves with, in a global, abstract, defining sense.

The Metaparadigm of Clinical Dietetics

Each concept in The Metaparadigm of Clinical Dietetics can also be referred to as a domain: one of the areas, or domains, of concern to the discipline. For example one domain of concern to clinical dietetics is the environment of the client. Another domain of interest or concern is the environment of the practitioner. Another domain includes the human conditions that clinical dietitians address in the practice of the profession: conditions of food allergies, disease, weight status, etc.

The seven domains of The Metaparadigm of Clinical Dietetics are:

Metaparadigm of Clinical Dietetics						
Reference Person	Human Condition	Practitioner Actions/ Attitudes	Practitioner Environment	Client Actions/ Attitudes	Client Environment	Nutraceuticals

Metaparadigm statements are profession-specific, ie, they are defined using specific terminology relevant to the profession. Definitions for domains of The Metaparadigm of Clinical Dietetics follow. Other professions may have structures for the body of knowledge they utilize and metaparadigms that define that profession's concerns in a similar global sense. The concepts, or domains, of The Metaparadigm of Nursing are: Person, Environment, Health and Nursing (Fawcett, 1984) (Smith, 1979).

The characterizations, or knowledge topics, describing each domain are also profession-specific. Ninety-four knowledge topics were selected to characterize, or describe, the seven domains of The Metaparadigm of Clinical Dietetics. As the body of knowledge expands and practice changes, the knowledge topic characterizations may change. It is possible that clinical dietetics may also change their domains of concern: Who knows what the future of clinical nutritional care holds? The Metaparadigm of Clinical Dietetics and the 94 knowledge topics can organize the "chaos" of knowledge used and describe it. This has the potential for facilitating the development of more theories, more science and more knowledge.

Definitions of the Seven Domains

<u>The Metaparadigm of Clinical Dietetics</u> refers to the seven global, abstract concepts that describe the phenomena of concern to the profession of clinical dietetics and are the guiding concepts which encompass the less abstract levels of professional knowledge. The definition of each domain, or concept of concern, is one way to define that discipline or profession and differentiate it from other groups. These other groups include, but are not limited to, other health professions or nutritional scientists.

<u>Reference Person:</u>   Reference Person refers to the theoretical, statistically derived individual representative of the reference population, for example the reference "infant 0.5-1.0 years old" referred to in the Recommended Dietary Allowances. It includes the assumption of a defined criteria of selection and assumes the user is informed regarding the essential details of the derivation. When reference values are used in evaluation or interpretation, it is acknowledged that health and disease are relative, not absolute states.

<u>Human Condition:</u>   For clinical dietitians Human Condition refers to the nutritional status of individuals in a state of health or with nutritional problems. The scientifically derived reference status is compared with observed departures from "Normal" status to assess the human condition of clients. Such assessment gives direction to the practitioner providing nutritional care to clients. Human Condition includes the Nutriologic Person, and PsychoNutriologic Person. NutrioGenetic Person is a possible addition to this domain.

<u>Practitioner Actions/Attitudes:</u>

<u>Actions</u> refers to behaviors engaged in or purposefully refrained from, relative to the practice of the profession or to professional development.

<u>Attitude</u> refers to intrapersonally-based thoughts and feelings about aspects of the clinical dietitian's professional role enactment that are elicited by situational cues. They may be explicit and willfully affect professional behavior, or implicit (unstated, unknown, subconscious or unconscious) and involuntarily affect behavior, not being under the influence of the will. In this instrument attitude includes, but is not limited to ability, aptitude, beliefs, decisions, emotions, ethics, ideas, knowledge, morals, opinions, preferences, thoughts, values, will and world view.

<u>Practitioner Environment:</u>   Practitioner environment refers to the complex social and physical circumstances in which clinical dietetics is practiced. It includes relationships with other professionals,

the local and national organization, the prevailing political and social milieu, the scientific knowledge, the state of technology available and profession-specific tools.

Client Actions/Attitudes:

Client refers to any individual or groups that present to a clinical dietitian for nutritional services.

Attitudes: refers to intrapersonal characteristics and processes of a client. Attitudes may be explicit and willfully affect client behavior or implicit (unstated, unknown, subconscious or unconscious) and involuntarily affect behavior, not being under the influence of the will. In this instrument attitude includes, but is not limited to ability, aptitude, beliefs, decisions, desires, emotions, ethics, ideas, knowledge, morals, opinions, preferences, thoughts, values, will and world view.

Client Environment: Client environment refers to the complex social and physical circumstances surrounding a client who receives nutritional interventions from a clinical dietitian. It includes influences of health status and medications, family and associates, food available, work, finances, cultural influences, the marketplace and self care skills.

Nutraceuticals: Nutraceuticals refers to any substance that can be considered to be a food or a component of a food that affects health, including the prevention and treatment of disease. Such products range from all natural, processed, created/engineered, manufactured foods, designer food, functional foods, phytochemicals, isolated nutrients supplements, chemopreventive agents and pharmafoods (Nutraceutical initiative, 1992), (Position Paper of ADA, 1995).

Paradigm of the Professions

The paradigm of the professions, as discussed by Wertheimer and Smith (Wertheimer and Smith, 1989) (Codes of Federal Register, 1991), while reflecting on the profession of pharmacy, includes a unique body of knowledge as one of the defining characteristics of a recognized profession.

Paradigm for the Professions
(a pattern of professional behavior)

1. An ideology—based on original faith professed by a profession.

2. An ethic—binding on the practitioners

3. A body of knowledge—unique to a given profession

4. A set of skills—when combined form the technique of a profession

5. A guild—of those entitled to practice a profession

6. Authority—granted by society (professional respect, licensure or certification)

7. An institutional setting—practice in a standardized environment

8. A theory—based on societal benefits derived from the ideology

Wertheimer AI, Smith, MC. Pharmacy Practice: Social and Behavioral Aspects. Baltimore. John P. Butler, editor, Third edition, Williams and Wilkins. Baltimore; 1989.

Clinical dietetics may be mapped onto the Paradigm of the Professions with reference to The Metaparadigm of Clinical Dietetics, including knowledge topics perceived by practitioners to be unique to the profession. See the following pattern:

Pattern of a Profession	Clinical Dietetics Characterization
1. An ideology	American Dietetic Association statements of philosophy and mission
2. An ethic	American Dietetic Association Code of Ethics
3. A body of knowledge unique to the profession	As defined by The Metaparadigm of Clinical Dietetics validation study; 27 of 94 knowledge topics were perceived as unique
4. A set of skills/techniques	*Assessment of* diet, environment, client knowledge, ability, preferences, food choices, food resources; social milieu *Provision of* nutritional interventions—menus, education, counseling, goal-setting, recommendations, treatment protocols *Expression of* findings using Nutrition Diagnostic codes *Documentation of*—activities, outcomes, management *Use of* available tools and technology, nutrition educational materials
5. A guild	American Dietetic Association; Registered and/or Licensed Dietitians, continuing education
6. Authority	Joint Commission for Accreditation of Hospitals standards, reimbursement for nutritional interventions, MNT
7. An institutional setting	A variety of institutions which provide health care and wellness maintenance
8. A theory	Eating a healthy variety of food supports health maintenance; medical nutrition therapy reduces risks associated with disease.

A Body of Knowledge Unique to the Profession

Since the founding of the American Dietetic Association (ADA) in 1925 (MacEachern, 1925) there has been no elucidation of the structure or definition of the general contents of the body of knowledge used in clinical dietetics. Changes in the health care system plus the increased interest in and knowledge about the influence of nutrition upon health (Blackburn, 1979) (Harper, 1991) have created a need for clinical dietitians to reflect upon their role and unique identity (DeBusk, 2000).

The potential contribution of clinical dietetics to health care in the United States can be enhanced by metaparadigm-driven dietetic health care, education of practitioners and research elucidating the phenomena of concern to the profession. Having in mind the definition of clinical dietetics and the profession's concerns will assist clinical dietitians in building the body of knowledge that is their unique contribution to health care.

The Metaparadigm of Clinical Dietetics and the seven domains can be utilized in defining the scope of practice of Clinical Dietetics. A defined and recognized scope of practice has been an integral part of licensure and third party payment for services by clinical dietitians. The initial proposal of a practitioner-validated Metaparadigm of Clinical Dietetics can present a global picture of what practitioners consider within their scope of concern when practicing clinical dietetics. It provides the theoretical underpinning and organization of knowledge applied while practicing clinical dietetics.

SECTION A

The Derivation of the Metaparadigm
of Clinical Dietetics

I. Metatheory: Derivation and Validation
of The Metaparadigm of Clinical Dietetics

Statement of Purpose

The purpose of the study was to propose and validate a Metaparadigm of Clinical Dietetics by soliciting perceptions of practitioners from selected practice groups in the American Dietetic Association. Perceptions were requested concerning relevance of knowledge topics derived from clinical dietetic and other literature to determine if practitioners perceived the proposed Metaparadigm of Clinical Dietetics as encompassing the profession's body of knowledge as characterized by ninety-four selected knowledge topics.

A further purpose was to seek the perceptions of practitioners concerning the comparative relevance of the same knowledge topics to nutritional scientists and/or to other health professionals. This would indicate which knowledge was perceived as shared and which knowledge was perceived by practitioners as unique to the profession.

History, Knowledge and Rationale for Study to Propose and Validate Metaparadigm of Clinical Dietetics

Background and Significance

Since the founding of the American Dietetic Association (ADA) in 1925 (MacEachern, 1925 there has been no elucidation of the structure of the body of knowledge used in clinical dietetics. A unique body of knowledge is one of the characteristics of a recognized profession (Wertheimer, Codes of Federal Regulations).

A metaparadigm is a statement of global, abstract, over-arching concepts that describe the phenomena of concern, or domains of relevance, of a profession. It represents the actual and/or potential reality for a discipline and acts as a framework that encompasses the less abstract levels of knowledge in the structural hierarchy of the body of knowledge (Leyse & Kight, 1993). It could be called a paradigm of paradigms. The metaparadigm is the abstract level of knowledge which will encompass increasingly concrete levels of knowledge. A metaparadigm has been further characterized as perspective-neutral and international in scope and substance (Fawcett 1984, 1995).

The proposed structure of clinical dietetics' body of knowledge is a hierarchy of concepts on a continuum from very abstract (the metaparadigm) to empirical (the science), not a hierarchy of value or truth. Within the hierarchy is the potential for knowledge-building with abductive processes, (Reed 1995) induction, deduction, also including reduction and incorporation methods (Rubenstein 1984).

Clinical dietetic practitioners need system theories and also theories concerning relationships with clients which encompass empathy, experience and subjectivity Systems sciences argue that hierarchy is essential for integration, wholeness and systems functioning … Hierarchy is simply a ranking of phenomena according to their holistic capacity … not necessarily their value, domination or oppression … Systems theories are essentially theories of surfaces or exteriors. To understand interiors (subjectivity, experience, consciousness) requires another approach, namely empathy, introspection and interpretation" (Walsh, 1985).

The non-Newtonian theories of quantum physics combined with chaos theory can add to understanding phenomena of concern to clinical dietetics (Margaret J Wheatley, 1999). The human body, its systems, the functions of individual nutrients, all operate as feedback systems that produce "new knowledge—physiologic knowledge", and functional adjustments over time. Functions vary according to the unique direction of each individual's genetic structure. The social/educational/counseling interaction of clinical dietitians and their clients are unique to each diad and undergoes change during treatment and progress, creating new knowledge that is integrated into the system of treatment. These systems and the outcomes of the interactions are not always predictable or understood. The areas of Nutriologic Person and PsychoNutriologic Person in the domain of Human Condition needs ongoing research.

Changes in the health care system and increased interest in the influence of nutrition upon health have created a need for clinical dietitians and others to reflect upon their identity and role. (Toddhunter, 1964) (Cooper, 1967) (Baird, 1984) (Forcier, 1987) (Koop, 1988 (National Nutrition Monitoring and Research Act, 1990) (Healthy People 2000, 1991) (ADA Role Delineation study 1992) (Bierman, 1992) (Committee on Clinical Practice Issues, 1995) (Coulston, 1992) (Food and Nutrition Board, 1994) (Insull, 1994) (Mason, 1994) (Parks, 1994) (The American Dietetic Association, 1995) (Brownell, 1995) (Covey, 1995) (Young, 1995).

Clinical dietetics has professional ties to nutritional science and to other health care professions. Therefore clarification of clinical dietetics' scope and boundaries involves defining which knowledge topics practitioners perceive as relevant to the practice of dietetics and distinguishing between knowledge topics perceived as unique and knowledge topics perceived as shared with nutritional sciences and other health professionals.

Knowledge in Nutritional Science

From the perspective of nutritional science, Harper states "Separation of disciplines is an artificial process, an organizational mechanism of the human mind to simplify the management of information ... yet there is a body of knowledge that is strictly nutrition (Harper, 1991)." Nutritional science defines empiric knowing as development of understanding and conclusions by the process of scientific observation. However, the classical school of empiricism that prevailed for fourteen centuries was based on observations alone, disregarding all theoretical and philosophical considerations ... giving rise to erroneous assertions and opinions as if they were fact. Experimentation and analysis were added to empiricism by medicine, nutrition and others to contribute to empirical knowing and the creation of scientific knowledge (Blackburn, 1979).

In solving nutritional problems, Schuftan admonishes readers that "all science should be a search for meaning. Nutritional scientists must recognize the interplay between knowledge and power and that intellectual development cannot be separated from moral development. In accepting the perception that food and nutrition interventions are intrinsically good, nutrition scientists have to realize the importance of programs' social and political context (Schuftan, 1987).

Nutritional science has historically worked within the modernist scientific paradigm, reducing the whole to smaller and smaller units, addressing only those units which can be physically observed and developing technology to measure ever smaller units with greater precision. Investigating the influence of nutritional factors on expression of single genes occupies many nutrition scientists (Simopoulos, 1995).

Knowledge in Clinical Dietetics

Clinical dietetics has historically concentrated on empirical knowledge, a legacy from nutritional sciences. Empirical knowing, the use of science and the scientific method, is generally regarded by clinical dietitians as the way to conceptualize knowledge that is authentic and credible. "Clinical dietetic research allows objective measurement of complex environments and tangible evaluation of the outcomes of procedures and treatments. The strength of a discipline ... is closely associated with its research base (Monsen, 1988)." Empirical research methods used by clinical dietitians include qualitative, case studies, surveys, experimental, quasi-experimental, cohort, and case-control. A combination of deductive, qualitative, and inductive methods were used in this study to define and validate the proposed Metaparadigm of Clinical Dietetics.

Levels of Knowledge

The Abstract Level of Knowledge:
Metaparadigm Theory Development

The proposed metaparadigm of clinical dietetics is grounded in Kight's research typology for client-centered dietetics which specified five domains (The Human Condition, Practitioner Actions, Practitioner Environment, Client Attitude, Client Environment) (Kight, 1986) paralleling the initial development in nursing. Reflection by an expert panel of clinical dietitians (N = 6), resulted in strong agreement that a metaparadigm is

1) representative of a discipline's uniqueness,

2) representative of actual/potential reality for a discipline,

3) made up of guiding units for building a discipline's structural hierarchy of knowledge,

4) general/global/highly abstract in nature,

5) an encapsulating unit/framework representative of a given discipline and

6) is made up of central units before/beyond paradigms, which are components in a structural hierarchy of knowledge (Leyse & Kight, 1993).

It was further agreed that domains encompassing the phenomena of concern to clinical dietitians are 1) **Reference Person,** 2) **Human Condition,** 3) **Practitioner Action/Attitude,** 4) **Practitioner Environment,** 5) **Client Action/Attitude,** 6) **Client Environment and** 7) **Nutraceuticals.**

The Second Level of Knowledge:
Paradigms, Conceptual Models, and World View

Paradigms

The relationships between domains of the proposed Metaparadigm of Clinical Dietetics will be demonstrated in the future in the paradigms, conceptual models and world views of practitioners as more conscious development of the body of knowledge progresses. Paradigms are not measurable because abstract concepts are used to express this level of knowledge. More concrete knowledge topics thought to be included and relevant to paradigms of clinical dietetics are included in the validation survey.

"Struggle with the concept (of paradigms) is evidence of coming of age—not proof of immaturity of a discipline. The scientific status of a discipline cannot be improved by legislating agreement on fundamentals and then turning to problem solving (Eckberg, 1979)." Progress in building science is made by sequential concrete solutions to solve puzzles or problems encountered and generating more puzzles. Paradigms are generated by the interaction of the level of analysis (micro/macro) and the substantive component emphasized in the question (material, affective or symbolic). Paradigms are often not discipline-wide, but are encountered in substantive areas of research (Eckberg). Paradigms are said to shift when they lose explanatory power. The protocols being tested for dietetic treatment of non-insulin-dependent diabetes, putting management of the disease in the hands of the client, has been identified as a paradigm shift away from the usual assumptions imbedded in the concept "diabetic diets" (Carey, 1995).

Conceptual Models

Clinical dietetics' paradigms and/or conceptual models are not often formalized, but likely exist due to the course content required by the ADA for educating dietitians. A conceptual model by G. Arroyave (in Young and Scrimshaw, 1979) still widely used in clinical dietetics and nutritional science, shows temporal relationships and development of clinically evident nutritional disease.

A refinement of the conceptual model of the Recommended Dietary Allowances, showing both deficiency and toxicity ranges, is anticipated to be more reflective of reality and knowledge, and therefore more explanatory, and more useful to the practitioner than the previous model showing only deficiency ranges (Food and Nutrition Board, 1994).

World Views

Any discussion of knowledge and levels of knowledge needs to recognize the world view from which the discussion is launched. One's professional and personal world view determine what is defined as knowledge, therefore influencing all else that follows. World view is the overall perspective or basic beliefs about what is real, what is true, the nature of human beings, how the world works, what is desirable, ethical. Philosophy is defined as the study of and statements regarding the values, beliefs and perspectives on the world, truth, reality, knowledge, purpose and existence. World view and philosophy are therefore similar concepts.

A world view that the whole can be created from component parts gives a different view of knowledge or truth than the world view that the whole is greater than the sum of the parts. Acceptance of only physically observable phenomena will preclude knowledge development regarding phenomena such as attitude, emotional stress, mental processes and group interaction. The world view and paradigms of clinical dietetics will appropriately include diverse phenomena. Clinical dietetics' world view will influence the process of educating practitioners as well as the content of the education.

The official world view of the American Dietetic Association is seen in the statements of mission, vision, philosophy and values. The contents were developed by consensus in the House of Delegates of the ADA (see appendix D).These statements by the American Dietetic Association guide the practice of clinical dietetics.

Personal world view cannot be divorced from practice and one's professional world view. However, the ADA has a censure procedure that may be exercised if a practitioner is found to be operating outside the values of the organization. The fact that it is seldom invoked could be construed to mean clinical dietitians are well enculturated during their training period, or possibly, the profession appeals to individuals with certain predictable characteristics.

Ethical issues are part of a professional's world view. In considering the ethics of withholding food and/or water in individual situations, practitioner actions are linked to the value systems of the patient, the patient's significant others, the health care team members, and the facility. This contextual world view is also evidenced by consideration of the political and legal ramifications in the development of decisions which may change or develop over time, with one patient or with successive patients (American Dietetic Association, 1995).

This investigator's perception is that the world view of clinical dietitians' includes, but is not limited to, the following assumptions: the clinical dietitian uses knowledge that is as scientifically and clinically correct as is possible at any one point in time, the client has ability to change, the environment influences the client and can be influenced by the client, and changes by the client as recommended by the clinical dietitian will have a positive outcome. "Positive outcome" is usually defined by the practitioner and/or health care system and generally includes symptom relief or disease risk reduction. It is most often assumed that the client shares these assumptions and definitions. These assumptions are usually not explicitly addressed, which demonstrates the influence of paradigms. Making paradigms explicit is one area of potential knowledge development for clinical dietetics.

The world view of clinical dietetics is influenced by the world view of nutritional science which includes study of the particulate while historically keeping in mind that a whole organism is influenced by the functioning of the smallest unit measurable at any one point in history (Toddhunter, 1967). "Advances in nutrition science will inevitably raise questions of values, which are para-scientific.… Is a population taller, and are they happier, or will they be productive in arts and sciences? … It is reasonable to envisage a time that dietary regimes can be designed for not only growth and health, but also for function and cultural values (Dubos, 1979)."

The Third Level of Knowledge:
Theories and Hypotheses

Measurable relationships among concepts are the next level in the hierarchy of clinical dietetics' knowledge. Hypothesis refers to a relational statement between two empirical indicators (Dulock, 1991). For example, in scientific research it could be a statement about a population characteristic that predicts the outcome of a study. Theory refers to an internally consistent group of relational statements that present a systematic view about a phenomena and that is useful for description, prediction and/or control (Walker & Avant, 1988). Theory provides a discipline with the means to articulate its focus (Meleis, 1991). Theories and hypotheses demonstrate absence/presence, association/correlation, cause/effect, predictive, or prescriptive relationships or temporal patterns of occurrence.

In the proposed organization of knowledge of clinical dietetics, hypotheses and theories operationalize and test the more abstract ideas found at the level of paradigms and/or conceptual models. At this level concepts are defined precisely, relationships are explicit, hypotheses are stated in measurable terms, methods of measurement are defined and degree of generalizability is acknowledged. The content of potential theories and hypotheses accepted in clinical dietetics are included in the survey validating the metaparadigm.

A statement by Dickoff and James on theories in nursing can be applied to clinical dietetics. "There is no question whether nursing (read "clinical dietetics") should or should not have theories … a professional discipline must provide for more than mere understanding; it must provide for conceptualization especially intended to guide the shaping of reality for actualizing the professional purpose (Dickoff & James, 1968)." A metaparadigm is a metatheory (a theory concerning theories) of clinical dietetics. It defines the basic professional concerns from which the profession's paradigms and theories are fashioned. A validated metaparadigm will provide such a conceptualization of the body of knowledge for clinical dietetics.

In comparison, nutritional science commonly deals with micro-theories, addressing phenomenon at the subsystem, molecular and genetic level. Nutritional scientists are also involved in anthropology (nutrition and culture), epidemiology (nutrition in social systems), food science (nutrition and food production), medicine (nutrition in diseases), or physiology (nutrition related to physical performance and/or body composition).

Clinical nutritional science addresses the whole person. It deals with nutrients, their mental effects and physical effects at the level of tissue, organs and body composition. It may include the functional capacity of the whole person. In addition it includes study of the effects of foods as consumed and effects from the environment, which may differ in effect from isolated nutrients. It also includes considering the meaning attributed to foods and eating circumstances.

The Fourth Level of Knowledge: Science

The need for imposing order on experience is a human need: order exists in the mind of the curious scientist-observer with perception and thought as the instruments used (Dubin, 1978). "Science is the systematic acquisition of knowledge derived from observation, experimentation, and analysis conducted to determine the nature and principles of what is being studied (Blackburn, 1979)." The metaparadigm concepts are the broad statement of concerns that clinical dietetics practitioners will observe, describe, experiment with and analyze.

Knowledge begins with the identification and naming of a phenomenon (object of sensuous perception) or noumenon (object of purely intellectual intuition). A name infers nuances, conjures images, provokes emotions, hints at political context; it can imply the paradigm of the one who names. "… researchers have a responsibility to recognize and consider the names which define the discipline (Muller, 1993)." "All human thought and inquiry is guided by some theoretical perspective, including the recognition and naming of phenomena.… A new and unique knowledge base does require a new way of speaking (Cody, 1992)." The new name and concept "Metaparadigm of Clinical Dietetics" is a new way of thinking for clinical dietitians.

Prior to and during the early 1900's, the identification and naming of nutrients was the mission of nutrition science. Nutritional science and food technology has recently spawned a host of new foods and terms still in the process of being defined and differentiated. The term "Nutraceutical" in the proposed Metaparadigm of Clinical Dietetics was selected from new terms (Nutraceutical Initiative,1992) in the food and health industry which includes phytochemicals, pharmafoods, functional foods and designer foods (see Glossary). ADA's position paper on nutraceuticals supports research regarding the health benefits and risks of these substances and commits dietetics to ensuring the public has accurate scientific information in this emerging field (American Dietetic Association, 1995).

In addition to identification and naming, knowledge at the "science" level includes systems of classification, definition and description. The taxonomies of diagnosis (Kight, 1993), etiologies, interventions (Brownell, 1995) (Tripp-Reimer, 1996) (North American Nursing Diagnosis Association, 1990) and outcomes of dietetic and nursing care are examples of building blocks for enlarging the science of the respective professions (McClosky & Bulecheck, 1992). Also basic to the science of clinical dietetics will be the continued organized observations and data collected by clinical dietitians as they practice the profession.

Relationships Between Clinical Dietetics, Nutritional Scientists, Other Health Professionals, and the Health Care Milieu

Defining Clinical Dietetics

In 1917 dietetics was defined by the ADA as the "science of nutrition and the art of feeding people" (Forcier, 1977). In the first issue of the Journal of the American Dietetic Association in 1925 the dietitian was described as "closely related to the medical aspect … an adjunct to the doctor in observation, diagnosis and treatment of a special group of diseases … with contact with the clinician and the laboratory worker through the more extensive work being carried on in blood chemistry and metabolism … responsible for a well balanced special or general diet for all the clients of the hospital … responsible for the teaching of student nurses … whose curriculum demands a standard of dietetics to be taught." Research in clinical nutrition was seen as an integral part of the profession (MacEachern, 1925).

The 1984 role delineation study of the ADA defined clinical dietitian as "a health care professional credentialed as a registered dietitian who affects the nutrition care of individuals and groups in health and illness. "The clinical dietitian provides nutrition assessment, planning, implementation (including education and referral), and evaluation services; provides consultation for food service to coordinate nutrition care services, manages departmental and personnel functions for nutrition care services; delineates and manages external influences on the delivery of nutrition care. The clinical dietitian educates and coordinates activities as a member of the health care team; maintains skill and knowledge in optimal nutrition care; and conducts applied research." (Baird, 1984). Additional functions include counseling clients regarding nutrient-drug interactions and utilizing knowledge of the psychological aspects of eating, nurturing and the psychological consequences of nutrient deficiencies.

Clinical dietetics has also been defined as the emerging epidemiological branch of nutritional sciences (Kight, 1995) (Frank-Sporer, 1996) and as intervention specialists (Insull, 1994). Clinical nutrition care giving and knowledge development by the advanced-level, primarily nutriologic practitioner is seen by Kight (Kight, 2001) as the coming development in Human Biomedical Nutrition with practitioners being identified as Biomedical Nutritionists.

The knowledge shared and relationships presently existing will be the springboard from which future relationships between clinical dietetics, other health professions and nutritional sciences will be launched. To further elucidate the present perceptions, this validation survey requested practitioners' responses regarding comparative relevance of knowledge topics that may be shared with other health professionals and with nutritional scientists.

Relationships: Nutritional Science

Nutritional Scientist refers to academic scientists who elucidate problems and solutions related to nutritional status, nutritional requirements, nutrient metabolism and/or biochemistry, molecular and genetic biology related to nutrition.

The science of nutrition developed in the twentieth century, following concepts linking diet and survival, linking diet with disease in ancient Greece and diet with longevity in the sixteenth century. (Toddhunter, 1967) Clinical dietetics is grounded in nutritional sciences, applying the discoveries and accepted facts of nutritional science in clinical practice. Drawing a distinction between nutrition and dietetics in England, McLaren defines nutrition as being "physiological process" and dietetics as "an endeavor that has goals that are social." "Dietetics and medicine are both science and some art: nutrition on its own is not a true discipline while dietetics is, since it has a clear concept of its identity and goals" (Mason, 1994) (McLaren, 1994).

Nutrition and food sciences have been described as facing an identity crisis because of their interdisciplinary nature (Thomas & Earl, 1994). Nutrition is the "sum of the process concerned with growth, maintenance, and repair of the living body as a whole or, studied in vitro, within cells, organs, entire animals or humans (Olsen, 1986)." Nutritional scientists see their role as dealing with the fundamental aspects of physiology and chemistry, biochemistry, molecular biology and nutritional influences on genetic transcription (Simopoulas, 1995), although some define nutrition science as the science of food and its relationship to health (Olsen, 1986). Nutritional sciences research has also included the health and feeding of whole populations and includes issues regarding food supply.

Nutritional scientists and clinical dietitians acknowledge the need for communication on research findings, the philosophy guiding research, societal needs, ethics in decision-making, research and reporting findings (Cousins, 1996). Communication with governmental bodies that create legislation, and set health-related priorities for society, and who make research funding decisions is essential. Public servants need to know the perspective of nutritional scientists. The public and the media also benefit from having interpretation of research from clinical dietitians and nutritional scientists (Walker, 1995).

Relationships: Other Health Professionals

Other Health Professional refers to any other health profession/professional about which respondent has knowledge or experience.

As health team members, clinical dietitians function in a milieu of shared professional knowledge. Other health professionals including nurses, pharmacists, physicians, psychologists and others often observe and address nutrition-related problems in treating clients. The roles and titles of professionals with special nutrition education has been debated (Committee on Clinical Practice Issues, 1995) (Dutra-de-Oliveira, 1995) (Halsted, 1995) (Insull, 1994) (Mason, 1994) (McLaren, 1994) (Young, 1995).

Clinical dietetics is historically linked to nursing: both have a common ancestor in Florence Nightingale. In the 1800's she modified the "Regular" diet, called it the "Extra" diet and noted that sicker soldiers recuperated more quickly with this dietary change…. "(Nightingale) laid an excellent foundation (for dietetics) in her continued emphasis on the importance of properly chosen and well prepared food as a factor in treatment of the sick" (Cooper, 1967). Nursing has been defined as "the diagnosis and treatment of human responses to actual or potential health problems (Meleis, 1991)." Nurses have nutrition-related nursing diagnostic codes and may assess functional status in regard to nutrition and feeding (Tripp-Reimer, 1996) (North American Nursing Diagnosis Association, 1990) (McClosky & Bulecheck, 1992).

Pharmacists are mandated by the Omnibus Reconciliation Act of 1990 to counsel patients regarding medications, which may include over-the-counter medications as well as prescriptions. Parenteral feedings, nutrient-drug interactions, weight reduction aids and vitamin-mineral supplements could potentially be included in their counseling responsibilities (Smith, 1991) (Fink, 1997).

The inclusion of nutrition in the education and practice of medicine has been the subject of much debate. Nutrition courses in the curriculum of medical schools has been legislated, although compliance has been slow (National Nutrition Monitoring and Related Research Act, 1990) (Halsted, 1995) (Young, 1995). The title of Physician Nutrition Specialist (PNS) has been defined, a role proposed and financial support is being sought. The role defines a collaborative, supportive and supervisory relationship with clinical dietitians (Committee on Clinical Practice Issues, 1995). The diagnostic and treatment functions appear to be of interest to Physician Nutrition Specialists, while the time-intensive education and supportive roles are designated to dietitians. "Nutrology" has been suggested as a specialty area for physicians (Dutra-de-Oliveira 1995). The term "public nutrition" has been proposed to conceptually group nutrition-related issues for populations as compared to clinical nutrition which deals with the individual. It is proposed this inclusive term could replace the terms public health and international nutrition (Mason, 1996). Some question "whether it is possible or wise to … advance the recognition of nutrition as a medical discipline (Halsted, 1995)." The medical International Classification of Diseases (ICD9) (ICD, 1977) diagnostic codes include codes to diagnose/rule out nutrient deficiency diseases and for diagnosing types of malnutrition.

Clients taking psychoactive drugs have potential two-way interactions between nutrients and drugs (Chicago Dietetic Association, 1996). The Diagnostic and Statistical Manual of Mental Disorders fourth edition (DSM IV) includes diagnoses for various forms of eating disorders (DSM,

1994). Psychologists may be faced with food/nutrition/nutrient/drug issues when treating clients with anorexia nervosa, bulimia, compulsive overeating, depression, alcohol and drug abuse as well as mental retardation.

Other health professionals such as social workers and physical therapists may include in their professional assessments and treatment recommendations factors concerning ability to eat and the procurement and preparation of food.

Relationship: Social and Health Care Milieu

Dietary factors are associated with five of the ten leading causes of death and contribute substantially to the preventable illness and premature death in the United States (Healthy People 2000, 1991). Nutrition is identified as one of the three root determinants of death and disability in the US (Koop, 1988) (Young, 1995). Demographic changes anticipated in the U. S. population, as described in the Pew Commission report (Pew Health Professions Commission, 1995), will make interventions related to the interaction of nutrition and chronic health problems increasingly important. The ongoing application of new knowledge in molecular biology and food technology to the clinical area will be required. The psychology of individual and social change will be relevant to the fields of nutrition and dietetics. Clinical dietetic research will be one aspect of demonstrating efficacy of clinical interventions and of promising health-and-diet-related hypotheses.

"Research in the area of clinical nutrition bridges the gap between basic science research and clinical dietetic practice. The practice of clinical dietetics … is based on clinical nutrition research." The Research Agenda for Dietetics Conference calls for practice research concerning clinical conditions such as cardiovascular disease, diabetes, cancer, obesity, eating disorders, renal failure, chronic intestinal disorders, acquired immunodeficiency syndrome, chronic neurologic disorders and acute metabolic stress. (Coulston and Rock, 1992). Non-clinical areas of research for advancing the health care of populations includes, but is not limited to, areas of nutrition policy, societal issues, environmental issues, education and competency of practitioners, and cost-benefit ratio of nutrition services and programs. In addition to treating health problems, the dietetic profession is also seen as "serving the public through the promotion of optimal nutrition, health and well-being" (American Dietetic Association Position Paper, 1995). Experimental research and use of placebo controls, which have been the 'gold standard' of research, in the future may be complemented with the consecutive patient questionnaire database due to ethical and financial issues (Pincus, 1997) (Rothman & Michels, 1994).

Some say the dominant world view, or paradigm, of health care is changing … "led by the people, not science or medicine" (Rubik, 1996). Alternative health care modalities often include nutrition and diet either self-prescribed or practitioner-prescribed. "… although the use of vitamin and mineral supplements falls within the biomedical model, use in mega-doses or for purposes not endorsed or accepted within biomedicine would be classified in the United States as Complementary or Alternative Medicine (Panel on Definition and Describing, 1997)." Clients of clinical dietitians often seek information and/or advice concerning alternative health literature and products.

Summary and Conclusion:
History and Rationale for Metaparadigm of Clinical Dietetics

The following factors revealed in the literature serve as the rationale for this work:

1. Clinical dietetics has not previously defined the structure or uniqueness of the body of knowledge used by practitioners.

2. Clinical dietetic practitioners are trained to apply nutritional sciences in the clinical setting as well as to incorporate knowledge from the social sciences.

3. Practitioners of clinical dietetics share a body of pre-clinical and clinical knowledge with other health care providers. The boundary between clinical nutrition care provided by clinical dietitians and by other health care professionals is ambiguous. This lack of clarity includes traditional and non-traditional health care.

4. Recognition of the influence of nutritional factors on health problems in the United States is increasing. The mission of the American Dietetic Association includes contributing to the improvement of public health.

5. Knowledge developed through profession-specific research, guided by scientific theory and metatheory, as demonstrated by the nursing profession, will enrich the potential of contribution clinical dietetics to public health.

6. The proposed Metaparadigm of Clinical Dietetics has potential for encompassing various types and forms of knowledge. New types and forms of profession-specific-knowledge, shared knowledge and transformed knowledge fall within the boundaries of the proposed Metaparadigm of Clinical Dietetics. This organization of knowledge can guide the practice of clinical dietetics, the education of clinical dietetic practitioners and the generation of profession-specific knowledge.

Research Design, Research Questions

Research Design

This study used a descriptive-correlational-exploratory factor analytical design. A concept analysis of the concept metaparadigm was conducted by an expert panel of clinical practitioners (N = 6). Deductive examination of dietetic literature by the expert panel and additional literature by the investigator was followed by a written survey of the expert panel's perceptions and level of agreement concerning the characteristics of a metaparadigm and the fit of literature-based knowledge topics into the seven proposed domains (Leyse and Kight, 1993). Structured qualitative interviews of six additional practitioners were conducted. Interviews were transcribed, comments were reduced to themes and compared to the knowledge topics derived from the literature and the expert panel's work. A pilot of the survey instrument was completed by four additional practitioners. These processes were followed by the inductive validation of the proposed metaparadigm using a written survey mailed to practitioners to determine perceptions of a random sample of practitioners. These processes will be described in detail in this paper.

Research Questions

Research Question 1:

Do clinical dietitians perceive the domains represented by the proposed Metaparadigm of Clinical Dietetics as encompassing the body of knowledge relevant to clinical dietetics?

Research Question 2:

Which knowledge topics are perceived as unique to clinical dietetics and which topics are perceived to be shared with selected other health professionals and/or nutritional scientists?

Other Health Professional refers to any other health profession/professional about which respondent has knowledge or experience.

Nutritional scientist refers to academic scientists who elucidate problems and solutions related to nutritional status, nutritional requirements, nutrient metabolism and/or biochemistry, molecular and genetic biology related to nutrition (Forman & Halsted, 1998).

Relevance refers to any phenomenon of concern that pertains to, is influential, important or applicable to either the practice of clinical dietetics, other health professionals or to nutritional science or is an integral part of the body of knowledge of each, respectively.

Research Method

Instrument Development and Testing

The survey instrument (see appendix) was constructed using the first four steps in classical test theory: the concept was defined, a self-report scale designed, items were reviewed, and a preliminary tryout of items conducted. This survey could be considered a field test according to classical test theory (Burns & Grove, 1993).

An expert panel of clinical practitioners educated in theory-building (N = 6) (Burns & Grove, p. 344) was convened at regular intervals for three months. Characteristics of a metaparadigm, Kight's previously published dietetic research typology (Kight, 1986) and all clinical articles from the *Journal of the American Dietetic Association* from 1978-1993 were reviewed and discussed. Topics were sorted and clustered to assess the fit of topics into Kight's five-domain typology and the additional domains of Reference Person and Nutraceuticals as well as the addition of Practitioner Attitudes and Client Attitudes to Practitioner and Client Actions were added. The investigator reduced the literature into knowledge topics and clustered the knowledge topics into the proposed domains. A written questionnaire was completed by members of the expert panel regarding characteristics of a metaparadigm, determining level of agreement and perceived fit of literature-based knowledge topics into the seven proposed domains. Results were published by Leyse and Kight (1993).

Structured, open-ended qualitative interviews of six additional practitioners were conducted to ascertain practitioner's perceptions of the phenomena of concern to the discipline, shared and unique concerns, and meanings of terminology under consideration (see Appendix C). Practitioners were clinical dietitians practicing in Tucson, from a variety of practice environments, (psychiatric inpatient, university medical center and HMO) with one to fifteen years in practice and educational levels from BS degree to doctoral student. Interviews were held in the practitioner's work environment and lasted approximately forty to sixty minutes. Interviews were taped and transcribed, transcripts were reduced to themes, then knowledge topics, and compared to the knowledge topics derived from the literature and the expert panel's work. The knowledge topics were theoretically placed into the domains of the proposed metaparadigm to determine applicability. The selection of a relevance scale was influenced by these interviews based on interviewee's perceptions of the term relevance compared to their perceptions of the term fundamental.

Additional literature was reviewed by the investigator: knowledge topics were reduced to ninety-four for the final survey. Forty-seven of the ninety-four knowledge topics were those considered by the expert panel. Other topics were derived from 1) abstraction of concrete topics into more general topics, 2) separation of complex concepts into elemental concepts to prevent confounding,

3) were topics the expert panel did not discuss sufficiently to give a clear level of agreement or 4) were added based on qualitative interviews or review of the literature.

Relevance Scores of Knowledge Topics

Perceptions of relevance and comparative relevance for ninety-four dependent variables were measured on a four-point Likert-like scale (Burns & Grove, 1993). A forced choice four-point scale was selected to eliminate a neutral response from respondents. Perceived relevance to clinical dietitians was indicated on the scale shown in the following table.

Scales for Measuring Relevance of Knowledge Topics to Clinical Dietetics and Comparative Relevance of Knowledge Topics to Other Health Professionals and Nutritional Scientists.

Relevance Scale	Numerical Score
Irrelevant (IR)	translated to score = 0
Somewhat Irrelevant (SIR)	translated to score = 1
Somewhat Relevant (SR)	translated to score = 2
Relevant (R)	translated to score = 3

Comparative Relevance Scale	Numerical Score
Not Relevant (NR)	translated to score = 0
Less Relevant than to Clinical Dietetics (LR)	translated to score = 1
Equally Relevant to Clinical Dietetics (ER)	translated to score = 2
More Relevant than to Clinical Dietetics (MR)	translated to score = 3

In addition to scoring the 94 knowledge topics for relevance and comparative relevance, respondents to the written survey were asked to specify any health professionals they considered rather than just the four initially specified in the survey (Nurses, Pharmacists, Physicians, Psychologists). They were also asked whether they felt the 94 knowledge topics were encompassed by the seven domains of The Metaparadigm of Clinical Dietetics. Respondents were asked to place their own work within the domains of the metaparadigm if possible, with multiple answers allowed. Demographic information was also requested from each respondent.

Pilot Test of Instrument

A late version of the survey instrument was pilot-tested by administering it to four clinical practitioners who were unfamiliar with any previous development. These practitioners were members of the Southern Arizona District Dietetic Association with varied levels of education and practice environments. The pilot test was conducted as a group activity in the conference room at the investigator's practice environment. The members of the group were provided only with the written instructions that would be mailed to sample respondents. The pilot was timed to determine what burden would be placed on those in the sample population. It took slightly less then one hour for these four practitioners to complete the survey. Therefore it was decided that the length of the survey was not an undue burden.

During the discussion following the administration of the survey it was determined that there was no misunderstanding of terminology and the survey was seen as requiring some thought, but not as an impossible task. Placement of instructions regarding "other health professionals" was changed at their suggestion.

The survey instrument was developed by specification on the computer and printed in the University of Arizona Testing Office with the assistance of Rick Haan, Computer Scientist to allow computerized data entry by scanning of completed surveys using an OPSCAN 5 automated scanner. The error rate of recording data for analysis is less than 0.5% with a potential of scanning 1000 pages per hour. The SAS software program and The University of Arizona mainframe computer were utilized for data analysis.

Reliability of Survey Instrument

Reliability (degree of internal consistency) of a measuring instrument (the written survey) is a prerequisite for assessing the validity of construct measurement. Measuring relevancy of knowledge topics is an indirect measure for the construct Metaparadigm. Testing reliability in measuring relevance or comparative relevance of knowledge topics is considered a measure of random error in the measurement technique. The magnitude of the reliability coefficient is directly related to the variance of the obtained scores. A Cronbach's alpha coefficient of 1.0 would represent perfect reliability. A level of $\alpha = 0.70$ was defined as acceptable for this new instrument (Nunnally & Bernstein, 1994) (Cody & Mitchell, 1992). The highest s.d. (standard deviation) for the 282 measures (3 professional groups X 94 knowledge topics) was 1.14. Four items had a mean s.d. of >1.0. All four were in the Comparative Relevance to Nutritional Scientists scores. The lowest s.d. was 0.086. Five items had a s.d. of, ≤ 0.20. All five were in the Relevance to Clinical dietetics scores. In general standard deviation decreased as relevance and/or comparative relevance increased. Assumptions of homoscedasticity (condition of equal variance) were therefore not warranted.

Cronbach Alpha Scores for Assessing Reliability of Survey Instrument, Calculated by Individual Domains

Domain	Professional group	Cronbach alpha Standardized variables
A. Reference Person	Clinical Dietetics	0.809
	Other Health Professionals	0.859
	Nutritional Scientists	0.886
B. Human Condition	Clinical Dietetics	0.743
	Other Health Professionals	0.807
	Nutritional Scientists	0.831
C. Practitioner Actions/Attitudes	Clinical Dietetics	0.855
	Other Health Professionals	0.869
	Nutritional Scientists	0.938
D. Practitioner Environment	Clinical Dietetics	0.853
	Other Health Professionals	0.869
	Nutritional Scientists	0.904
E. Client Actions/Attitudes	Clinical Dietetics	0.759
	Other Health Professionals	0.690 *
	Nutritional Scientists	0.899
F. Client Environment	Clinical Dietetics	0.692 *
	Other Health Professionals	0.773
	Nutritional Scientists	0.880
G. Nutraceuticals	Clinical Dietetics	0.847
	Other Health Professionals	0.845
	Nutritional Scientists	0.897

* These items do not meet the criteria of a Cronbach alpha value of $\alpha = 0.70$.

Two professional group scales were just under the critical value of $\alpha = 0.70$. One in the domain Client Environment for Clinical Dietetics, had a value of $\alpha = 0.692$. The other in the domain of Client Actions/Attitudes for Other Health Professionals, had a value of $\alpha = 0.690$. With rounding, they both met the criteria of 0.70 and were accepted.

Human Subjects Review

A description of this study and a copy of the survey instrument was reviewed and approved by the Human Subjects Committee of The University of Arizona. Participants in the interviews, pilot and written survey were treated anonymously. Names of the expert panel members were published in the report of their work and in the acknowledgments in this dissertation.

Data Collection Procedures—Sample and Sampling

Clinical Dietitian was operationally defined as a member of the American Dietetics Association who has passed the registration examination and has specified a clinical area of interest (a Dietetic Practice Group) on membership data forms. A sample of five hundred practitioners was selected based on a goal of responses equal to 2.5 times the number of dependent variables (2.5 X 94 = 235) for an optimum factor analysis and on previous response rates to surveys of clinical dietitians of >55% (Christie & Kight, 1993) (Thomson & Kight, 1990). With this response rate a final sample of 275 was anticipated. A random sample of five hundred members of thirteen selected Dietetic Practice Groups (DPG) was purchased from the American Dietetic Association in the form of four duplicate sets of mailing labels. The Dietetic Practice Groups were selected to encompass a variety of clinical specialty groups in addition to the general clinical practice group. The selection process potentially included practitioners from all states in the United States and any foreign members of the dietetic practice groups. There were no duplicate names in the sample obtained, even though some clinical dietitians belong to more than one dietetic practice group.

The sample drawn included three members from Canada, one from England and one from Puerto Rico. These five members were excluded from the mailings resulting in a final sample number of 495. In May, 1996 there were 29,440 members in the Dietetic Practice Groups selected. The sample of 500 represents 1.7 % of this population.

Selected American Dietetic Association Dietetic Practice Groups Represented in Study Sample and Percent of Members From Which They were Drawn

Dietetic Practice Group May 1996 membership		
Dietetic Practice Group (DPG)	No. of members	% of T DPG Members
DPG 10—Public Health Nutrition (PHN)	1300	4 %
DPG 11—Gerontological Nutritionists (GN)	2510	9
DPG 12—Dietetics in Developmental and Psychiatric Disorders (DDPD)	1170	4
DPG 20—Oncology Nutrition (ON)	930	3
DPG 21—Renal Dietitians (RPG)	1840	6
DPG 22—Pediatric Nutrition (PNPG)	2020	7
DPG 23—Diabetes Care and Education (DC)	3570	12
DPG 24—Dietitians in Nutrition Support (NS)	3530	12
DPG 25—Dietetics in Physical Medicine and Rehabilitation (DPMR)	460	2
DPG 27—Dietitians in General Clinical Practice (DGCP)	1950	7
DPG 28—Perinatal Nutrition (PN)	590	2
DPG 31—Consultant Dietitians in Health Care Facilities (CDHCF)	5400	18
DPG 33—Sports, Cardiovascular and Wellness Nutritionists (SCAN)	4170	14
Total membership in these practice groups	29,440	100 %

A record system was begun for tracking the date of each mailing and each reply received for each member of the sample. A letter was mailed to each practitioner selected in the random sample except those outside the U.S. The letter gave a brief rationale for the study, defined a metaparadigm and requested a commitment of one hour for completing the survey. A pre-printed postage-paid postcard coded with respondent's identification number was enclosed with the letter. Members of the sample were asked to return the postcard if they were willing to complete the survey. Each respondent was mailed an overview of the survey, a copy of the survey, a page of term definitions, instructions for completing the survey and a postage-paid, pre-addressed envelope for returning the completed survey. If the survey was not returned by eight weeks, a different colored reminder card was sent requesting they complete and return the survey. Ten weeks after the first letter was mailed, a second postcard was mailed asking those who had initially not committed to completing a survey to reconsider and return the second postage-paid postcard (Dillman, 1978) (Salant and Dillman, 1994).

Data Analysis and Results

Response Rate

The initial letter asking for a commitment of time and response to the survey was mailed to 495 clinical dietitians. Completed surveys were obtained from one hundred thirty-six respondents for a response rate of twenty-seven percent (136/495 = 0.27).

Missing Data

Missing relevance/comparative relevance scores were replaced with the mean score of that variable to prevent list-wise and pair-wise deletion during analysis. Frequencies and other descriptions were calculated omitting respondents with missing data. Missing data fell within a range of 0-6%. The extent of missing data is indicated in individual item results.

One item, Nutritional Counseling in the Clinical Dietetic section of the survey, had almost no variance. One hundred thirty-five respondents scored it as three on the relevance scale. One respondent scored it as two.

Description of the Sample and Respondents

Typical Respondent

Based on frequencies and modes, the typical respondent was female, lived in Ohio, New York, Florida, California or Virginia and belonged to one of the following four Dietetic Practice Groups: General Clinical Practice, Consultant in Health Care Facility, Gerontological Nutritionists or Public Health Nutrition. She was either 30-39 years old or 50-59 years old (bi-modal) and had been in practice 11-20 years. Her practice environment was a hospital, and she was nearly as likely to have a Master's degree as a Bachelor's degree (bi-modal). Her route of entry into the profession was through an internship and she was not likely to have been involved in research or to have authored publications.

Dietetic Practice Groups

Sixty-one percent of respondents represented four Dietetic Practice Groups (General Clinical Practice, Consultant in Health Care Facility, Gerontological Nutritionists or Public Health Nutrition). All thirteen selected Dietetic Practice Groups were represented by at least one respondent. "Total DPG Membership" proportions in the random sample were non-determinable. Clinical dietitians do not always become a member of the General Clinical Practice DPG, but rather, join the specialty group that corresponds to their work environment or interests. Some join several practice groups to keep current in several areas of dietetics. There was no duplication of names in the random sample. Ninety-four percent of respondents indicated their DPG, a missing data rate of six percent.

Comparison of Percent Survey Respondents and Percent of ADA Membership in Selected Dietetic Practice Groups

Frequency	% of Respondents	% Total ADA DPG Membership*
26 from Dietitians in General Clinical Practice DPG	20	7
19 from Consultant Dietitians in Health Care Facilities DPG	15	18
19 from Gerontological Nutritionists DPG	15	9
14 from Public Health Nutrition DPG	11	4
11 from Diabetes Care and Education DPG	7	12
8 from Pediatric Nutrition DPG	6	7
8 from Sports, Cardiovascular &Wellness DPG	6	14
6 from Renal Dietitians DPG	5	6
6 from Dietitians in Nutrition Support DPG	5	12
4 from Dietetics in Developmental and Psychiatric Disorders	3	5
3 from Oncology Nutrition DPG	2	9
2 from Dietetics in Physical Medicine &Rehabilitation DPG	2	2
1 from Perinatal DPG	1	2
Total 127	100	100
Missing Data	7	5.8

State of Residence of Respondents

Frequencies of practitioners from each of the fifty states in the original random sample and sample of respondents who completed a survey are compared below. See also the bar graph generated for a visual comparison.

In the initial sample one state was not represented. In the respondent sample fourteen states were not represented. States with the most respondents were New York, Ohio, Florida, California, Texas, Virginia, Massachusetts and Minnesota. These eight states accounted for 67 responses (49%). Northwest states (Washington, Oregon, Idaho, Wyoming, and Montana) and southwest states (New Mexico, Arizona, Utah, Nevada, Colorado, Oklahoma) were minimally represented in the survey. There were three respondents from the northwest states and four from the southwest states. Together these two areas accounted for 5% of the responses.

Comparison of Frequency Distribution of States of Residency of Practitioners from Initial Sample and Practitioners Who Completed a Survey.

State of Residence	Code		No. in Original Sample	No. of Respondents
Alabama	AL	01	6	2
Alaska	AK	02	3	2
Arizona	AZ	03	3	0
Arkansas	**AR**	**04**	**4**	**1**
California	**CA**	**05**	**56**	**8**
Colorado	**CO**	**06**	**7**	**3**
Connecticut	CT	07	10	1
Delaware	DE	08	3	0
District of Columbia	DC	09	1	0
Florida	**FL**	**10**	**24**	**9**
Georgia	**GA**	**11**	**11**	**2**
Hawaii	**HI**	**12**	**3**	**0**
Idaho	ID	13	3	1
Illinois	IL	14	28	4
Indiana	IN	15	13	5
Iowa	**IA**	**16**	**9**	**2**
Kansas	**KS**	**17**	**5**	**3**
Kentucky	**KY**	**18**	**4**	**0**
Louisiana	LA	19	4	1
Maine	ME	20	3	0
Maryland	MD	21	12	3
Massachusetts	**MA**	**22**	**11**	**7**

Comparison of Frequency Distribution of States of Residency of Practitioners from Initial Sample and Practitioners Who Completed a Survey.

State of Residence	Code		No. in Original Sample	No. of Respondents
Michigan	**MI**	**23**	**20**	**3**
Minnesota	**MN**	**24**	**16**	**6**
Mississippi	MS	25	8	3
Missouri	MO	26	14	5
Montana	MT	27	0	0
Nebraska	**NE**	**28**	**1**	**0**
Nevada	**NV**	**29**	**4**	**0**
New Hampshire	**NH**	**30**	**1**	**1**
New Jersey	NJ	31	9	3
New Mexico	NM	32	1	0
New York	NY	33	27	10
North Carolina	**NC**	**34**	**7**	**4**
North Dakota	**ND**	**35**	**3**	**1**
Ohio	**OH**	**36**	**35**	**10**
Oklahoma	OK	37	6	1
Oregon	OR	38	5	0
Pennsylvania	PA	39	23	5
Rhode Island	**RI**	**40**	**3**	**1**
South Carolina	**SC**	**41**	**10**	**5**
South Dakota	**SD**	**42**	**1**	**1**
Tennessee	TN	43	5	1
Texas	TX	44	30	9
Utah	UT	45	1	0
Vermont	**VT**	**46**	**1**	**0**
Virginia	**VA**	**47**	**18**	**8**

Comparison of Frequency Distribution of States of Residency of Practitioners from Initial Sample and Practitioners Who Completed a Survey.

State of Residence	Code		No. in Original Sample	No. of Respondents
Washington	**WA**	**48**	**13**	**1**
West Virginia	WV	49	2	1
Wisconsin	WI	50	11	3
Wyoming	WY	51	1	0
Missing data			1	
			N = 495	136

Number of Years in Practice

Eighty-one percent of respondents had been in practice more than ten years, the length of time it takes to become an expert in a field of endeavor. (Ericsson, 1993) Nearly half of the respondents had been in practice more than twenty years. There were no respondents who had entered the practice of clinical dietetics less than five years ago.

Number of Years Respondents have been in Practice in Clinical Dietetics

Years	Frequency	Percent
0-5	0	0.0
1-5	6	4.5
6-10	20	14.9
11-20	43	32.1
>20	65	48.5
Total	134	100.0
Missing data:	2	1.4

Environment in Which the Respondents Practiced Dietetics

In addition to the four categories provided, opportunity was provided in the survey for respondents to write in additional practice environments. Added written responses under "Other" on the survey were collapsed into two additional categories, the remaining responses making up the "other" category of environment. The added practice environments were Long Term Care and Out-Patient practice environments. The "other" practice environments included clinical research, health care consultant, public health, or a university, with three or four respondents in each category. Ninety percent of practitioners were employed in seven selected environments. Ten percent were in "other" environments. Fifty-seven percent of respondents were working in a hospital or long term care facility. Thirty-three percent were working with clients in the environments of out-patient clinics, WIC clinics, private practice or Health Maintenance Organizations.

Practice Environments of Survey Respondents

Environment	Frequency	Percent
Hospital	60	45.5
HMO	3	2.3
Private Practice	19	14.4
WIC(Women, Infants and Children)	6	4.5
Long Term Care	15	11.4
Out-Patient	15	11.4
Other	14	10.6
Total	132	100.1
Missing data:	4	2.9

Level of Education as Measured by Highest Degree Obtained

Forty nine and a half percent of respondents had obtained undergraduate degrees, forty-five percent had Masters degrees, five and a half percent had a doctorate degree. One hundred nine respondents (78%) had obtained degrees in science: B.S., M.S., and D.Sc.

Highest Educational Degree Obtained by Respondents

Highest Degree	Frequency	Percent
Bachelor of Arts	5	3.9
Bachelor of Science	59	45.7
Master of Arts	7	5.4
Masters of Science	48	37.2
Master of Public Health	3	2.3
Doctor of Philosophy	5	3.9
Doctor of Science	2	1.6
Total	129	100.0
Missing data	7	5.1

Route of Entry into Dietetics

Sixty-nine percent of respondents entered the profession via an internship. Fourteen percent entered the profession by the experience route. Fifteen percent entered the profession via the Coordinated Undergraduate Program (CUP) route. The experience routes were discontinued as an option for entering the profession in 1987 (Personal communication, Credentialing Division, American Dietetic Association). Therefore respondents taking this route entered the profession at least nine years ago. Two respondents entered through the Approved Pre-Professional Practice Program (AP4) route of entry.

Respondent's Route of Entry into the Profession

Route of Entry	Frequency	Percent
Internship	90	69.2
CUP (Coordinated Undergraduate Program)	20	15.4
Experience	18	13.8
AP4 (Approved Pre-Professional Practice Program)	2	1.5
Total	130	99.9
Missing data	6	4.41

Level of Professional Activity

Professional activity level was measured by four demographic items: 1) history of publications, 2) number of publications, 3) involvement in clinical dietetic research and 4) whether this research involvement was in the past or present (106). Twenty-six percent of respondents had published between one and six publications. Seven percent had published seven or more publications. Forty-six percent of respondents had been involved in research; one third were presently involved in research. Some respondents were involved in past research and continue research involvement.

Level of Professional Activity as Measured by Authorship of Publications

Publications	Frequency	Percent
No	90	66.7
Yes	45	33.3
Total	135	100.0
Missing Data	1	.7

Respondent's Involvement in Research

Involvement in Research	Frequency	Percent
No	73	54.1
Yes	62	45.9
Total	135	100.0
Missing data	1	0.7
Past research	25	18.4
Present research	46	33.8
Total	71	52.2

Age of Respondents

Sixty percent of respondents were fifty years old or greater. One respondent was less than thirty years old. Seventy-eight percent were between the ages of thirty and sixty.

Age of respondents

Age	Frequency	Percent
20-29	1	0.7
30-39	45	33.6
40-49	10	7.5
50-59	49	36.6
60-69	22	16.4
70-79	6	4.5
>79	1	0.7
Total	134	
Missing data	2	1.4

Gender of Respondents

Ninety-seven percent (N = 131) of the respondents were female, three percent (N = 4) were male. One person did not indicate gender. Gender-neutral names in the original sample prevented determining the proportion of females to males in the random sample obtained.

Respondents Requesting Results of the Survey

One hundred-eight (82%) respondents requested information regarding the results on the survey. Twenty-three (17.6%) respondents indicated they did not wish to be sent results of the survey. Five respondents did not indicate their preference.

Research Questions Answered

The following section addresses the research questions in order of presentation. The details of analysis and results are included in context of the research question.

Research Question 1

Do clinical dietitians perceive the domains represented by the proposed Metaparadigm of Clinical Dietetics as encompassing the body of knowledge relevant to clinical dietetics?

Relevance of Knowledge Topics

Ordinal data of perceived relevance was transformed to a numerical score and data was treated as interval data. Using the relevance scale of 0-3, knowledge topics with mean **scores of 2.5-3.0** were judged to be perceived as **Relevant** to Clinical Dietetics by respondents. Knowledge topics with mean scores of **1.5-2.49** were judged to be **Somewhat Relevant** to Clinical Dietetics. Knowledge topics with mean scores of **0.5 to 1.49** were judged to be **Somewhat Irrelevant** to Clinical Dietetics. Knowledge topics with a mean score of **0-0.49** were judged to be **Irrelevant** to Clinical Dietetics (see Appendix I). Ninety-nine percent of the knowledge topics were either Relevant or Somewhat Relevant to Clinical Dietetics, which yields a negatively skewed, leptokurtic distribution.

Relevance Scale

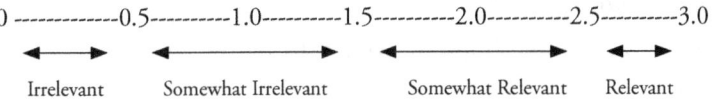

Knowledge Topics Perceived as Relevant to Clinical Dietetics in Descending Order Of Mean Relevance Score

Relevant knowledge topics with means between 2.90-2.99: N = 32 (34% of total)

Nutritional counseling

Nutritional recommendations *

Assessment of diet (eating habits, food preferences) *

Nutritional education *

Nutritional goal-setting *

Clinical dietetic continuing education *

Client having a care-giver who is responsible and knowledgeable re: nutrition needs assessment of environmental influences on clients (cultural influences, food availability, preparation facilities, social support) *

Dietitian—Client relationship *

Food choices made by clients

Route of food delivery (enteral/parenteral)

Acceptable laboratory test ranges

Nutrient functions

Nutrition education materials *

Assessment of individual's health status (medical dx, Rx, health/dental problems, mobility, vision, biochemical)

Signs/symptoms/potential for disease development related to nutritional status *

Client food sources *

Documentation of clinical dietetic activities *

Measurement of dietetic outcomes

Assessment of nutritional status (dietetic-specific physical examination, anthropometric) *

Level of client knowledge for nutritional self-care

Adequacy of finances of client for food *

Ability of client for nutritional self-care

Interaction of lifestyle status, health status with nutritional status

Food preferences

Knowledge Topics Perceived as Relevant to Clinical Dietetics in Descending Order Of Mean Relevance Score

Individual's food allergies/sensitivities

Use of dietetic treatment protocols

Nutrient disposition (absorption, utilization, excretion)

Registered status with ADA *

Recommended Dietary Allowances *

Misleading nutritional claims

Evaluation (of clients) for nutritionally treatable etiologies *

Relevant knowledge topics with means between 2.80-2.89: N = 13 (14% of total)

Assessment of individual's mental status (readiness to change, educational level, cognitive functioning) *

Potential or presence of nutrient/food/drug interactions

Knowledge of pre-clinical sciences: physiology

Client lifestyle choices made *

Influence of emotional status on food intake *

Tools for physical nutritional assessment of individuals *

Social milieu for eating *

Practitioner's personal professional philosophy

Nutritional reference values

Reimbursement for dietetic interventions

ADA code of ethics *

Effects of nutraceuticals

Determination of client transportation to acquire food *

Relevant knowledge topics with means of 2.70-2.79: N = 15 (15% of total)

Knowledge of anthropometric measurement *

Knowledge re: normal human aging *

Tools for assessment of nutrition-related behaviors of individuals *

Knowledge Topics Perceived as Relevant to Clinical Dietetics in Descending Order Of Mean Relevance Score

Prognostic Nutrition Indices *

Food-borne environmental toxins

Composition of nutraceuticals

Professional network, relationships

Knowledge of biochemistry

Functions of nutraceuticals

Normal growth patterns *

Management of clinical dietetic services *

JCAHO nutrition standards * (107)

Knowledge of integrated metabolism

Computer programs re: nutrition and/or diet *

Diagnostic thought processes

Relevant knowledge topics with means of 2.60-2.69: N = 8 (9%)

Public policy/law *

Changes in tissue appearance related to nutritional deficiency *

Knowledge of biology

Interdependent colleague relationships

State licensure *

Knowledge of psychology

Use of nutritional codes *

Intellectual multi-skilling (in-depth, science-based clinical practice, critical/global
    thinking, conceptualizing)

Relevant knowledge topics with means of 2.50-2.59: N = 9 (9% of total)

Tools for attitude assessment of individuals *

Menu writing *

Expression of findings using nutritional diagnostic codes, classification systems *

Knowledge Topics Perceived as Relevant to Clinical Dietetics in Descending Order Of Mean Relevance Score

ADA philosophy *

Electronic communication capability *

ADA mission statement *

ADA values statement *

Collaborative research

Funding for clinical dietetic research

Somewhat Relevant Knowledge topics with means of 1.50-2.49: N = 16 (16% of total)

Transformational leadership

Links with business/industry

Normal appearance of tissues likely to develop nutrient-based lesions

Client world view re: health and nutrition

Normal reproduction stages *

Nutritional factors in hormone regulation and gene expression

Knowledge of epidemiology

Board certified specialists—ADA *

Processing of nutraceuticals

Role of principal investigator in research

Theory-building in clinical dietetics *

Nutritional science laboratory methods *

Clinical Nutrition Specialist credential—American Board of Nutrition, AIN/ASNS

Statistical methods *

Fellow Status—ADA

Knowledge of anthropology

Somewhat Irrelevant Knowledge topics with mean of 0.50-1.49: N = 1 (<1% of total)

Technical multi-skilling (draw blood, take B/P)

No knowledge topics had a mean relevance to clinical dietetics score of less than 0.5

* = themes the original expert panel agreed or strongly agreed were clearly and convincingly representative of domains in which they were placed (N = 47). (1)

Missing data for knowledge topics ranged from 0-7 responses, a rate of ≤ 5 %.

Summary of Knowledge Topic Relevance to Clinical Dietetics

Relevance Score	Frequency	Percent
Relevant	77	82
Somewhat Relevant	16	17
Somewhat Irrelevant	1	1
Irrelevant	0	0
Total	94	100

Perceptions of the Proposed Metaparadigm and the Body of Knowledge

Ninety-five percent of respondents (N = 123) perceived the proposed Metaparadigm of Clinical Dietetics as encompassing the body of knowledge utilized in clinical dietetics. Five (4.7) percent (N = 6) did not perceive the proposed Metaparadigm of Clinical Dietetics as encompassing the body of knowledge. Topics mentioned as missing by respondents included some the investigator felt were covered in the survey but possibly overlooked, forgotten after completing the whole survey or perceived differently by the respondents than by the investigator. Topics listed by respondents were basic science, biology, budgets, business mentality, continuing competence, food, food composition and preparation, human and social psychology, pediatrics and public relations. Sociology and quality assurance was mentioned by one respondent and were not mentioned in the survey. Another respondent mentioned marketing the professional and skill in convincing other health professionals of our expertise. Seven respondents did not answer this question, a level of five percent missing data.

Respondent's Placement of Their Work Within the Domains of Metaparadigm

Respondents were asked to indicate into which domains of the proposed metaparadigm they placed their own work. Possible responses were from zero to seven domains. Six domains were selected as areas in which over fifty percent of responding practitioners placed their own work. The domain of Nutraceuticals was selected by forty-two percent of respondents. Sixty-five percent of respondents placed their work in four or more domains of the metaparadigm. A summary of responses is found in Tables 17 and 18.

Application of Metaparadigm by Practitioners—I

Domain	Frequency selecting	% selecting
Practitioner Actions/Attitudes	106	78
Client Action/Attitude	106	78
Client Environment	90	66
Human Condition	89	65
Reference Person	81	60
Practitioner Environment	79	58
Nutraceuticals	57	42

Application of Metaparadigm by Practitioners—II

Number of Domains in Which Respondents Placed Their Work	Frequency	Percent
0	5	3
1	16	12
2	15	11
3	12	9
4	9	7
5	26	19
6	31	23
7	22	16
Totals	136	100

Research Question 2

Which knowledge topics are perceived as unique to clinical dietetics and which topics are perceived to be shared with selected other health professionals and/or nutritional scientists?

The Scale of Comparative Relevance to Clinical Dietetics (CD) and Comparative Relevance scores of the Knowledge Topics were used to derive "Uniqueness".

Scale of Comparative Relevance to Clinical Dietetics (CD)

0 ------------0.5----------1.0----------1.5----------2.0----------2.5----------3.0

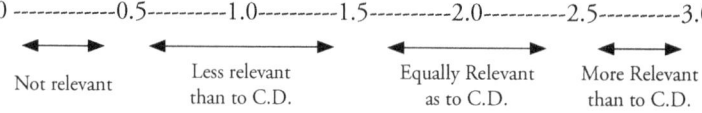

Not relevant   Less relevant   Equally Relevant   More Relevant
                than to C.D.   as to C.D.   than to C.D.

Knowledge Topics Perceived to be Unique to Clinical Dietetic

Uniqueness was determined by comparing the scores of Relevance to Clinical Dietetics with the scores of Comparative Relevance to Other Health Professionals and Comparative Relevance to Nutritional Scientists. Mean scores of items judged as Relevant to Clinical Dietetics (≥ 2.5) were compared to mean scores of each other group separately. Items perceived as having less relevance (a comparative relevance score of 0.5-1.49) or as having no relevance (a comparative relevance score of 0-0.49) to another group were judged to be unique to clinical dietetics. Items Relevant to Clinical Dietetics with a Comparative Relevance score of more (≥ 2.5) or equal relevance (1.5-2.49) to an other group were considered shared and not unique.

Example 1:

Nutritional Counseling knowledge topic (domain of Practitioner Actions/Attitudes):

Mean relevance score for clinical dietetic	= 2.999
Mean comparative relevance score for other health professionals	= 0.937
Mean comparative relevance score for nutritional scientists	= 0.865

Nutritional counseling was perceived to be unique to clinical dietetics.

Example 2:

Nutrient disposition knowledge topic (domain of Reference Person)

Mean relevance score for clinical dietetics	= 2.909
Mean comparative relevance score for other health professionals	= 1.453
Mean comparative relevance score for nutritional scientists	= 2.527

Nutritional disposition was perceived to be shared with nutritional scientists, with this topic being even more important to nutritional scientists and less important to other health professionals than to clinical dietetics.

Listed below are knowledge topics Relevant to Clinical Dietetics (relevance score ≥ 2.5) and less or not relevant to other health professionals or nutritional scientists, (comparative relevance score < 1.5) and therefore considered unique to clinical dietetics (N = 27). No knowledge topics from the Domains of Reference Person or Nutraceuticals met this criteria.

Unique Knowledge Topics Derived from Relevance Scores of Clinical Dietetics and Comparative Relevance Scores of Other Health Professionals and Nutritional Scientists.

Domain of Human Condition

    Nutritional Status

        a. Expression of findings using nutrition diagnostic codes, classification systems

Domain of Practitioner Actions/Attitudes

    Assessment of Individuals

        a. Diet (eating habits, food preferences)

        b. Environment (cultural influences, food availability, food preparation facilities, social support)

    Nutritional Diagnosis

        c. Use of nutrition diagnostic codes

    Provision of Nutritional Interventions

        a   Menu writing

        b. Nutrition education

        c. Nutrition counseling

        d. Nutritional goal-setting

        e. Nutritional recommendations

        f. Use of dietetic treatment protocols

    Documentation of Activities

        a. Documentation of clinical dietetic activities

        b. Measurement of dietetic outcomes

        c. Management of clinical dietetic services

    Philosophy of Clinical Dietetics

        b. American Dietetic Association philosophy

Unique Knowledge Topics Derived from Relevance Scores of Clinical Dietetics and Comparative Relevance Scores of Other Health Professionals and Nutritional Scientists.

Domain of Practitioner Environment

    Professional Credentialing

        a. Registered status-ADA

        d. RD Licensure—State

        f. Clinical Dietetic continuing education

    Professional Self-Defined Responsibility

        a. ADA Code of ethics

        b. ADA Mission statement

    Political/Social Milieu

        a. JCAHO nutrition standards

        c. Reimbursement for dietetic interventions

    Available Tools and Technology

        f. Nutrition education materials

Domain of Client Actions/Attitudes

    Acceptance of Responsibility for Self Care

        b. Knowledge for nutritional self care

    Influences on Food Intake

        a. Food Preferences

    Choices Made

        b. Food Choices

Domain of Client Environment

    Food Resources

        a. Food Sources

    Personal Social Resources

        c. Social milieu for eating

Knowledge Topics Perceived to be Shared

Listed below are knowledge topics Relevant to Clinical Dietetics (relevance score ≥ 2.5) and with a score of more or equal relevance to an other group (Comparative Relevance score > 1.5), and therefore considered to be shared knowledge (N = 49).

Knowledge Topics Considered be Shared Knowledge Derived from Scores of Relevance to Clinical Dietetics (≥ 2.5) and Comparative Relevance to Other Health Professionals and Nutritional Scientists (> 1.5)

Knowledge Topic and Domain	Comparative Relevance Scores Knowledge Shared With	
	OHP*	NS**
Domain of Reference Person		
Normal Human Functioning		
a. Nutrient disposition (absorption, utilization, excretion)		2.511
d. Integrated Metabolism		2.641
Normal Human Developmental Stages		
a. Aging	2.076	2.282
b. Growth patterns	2.060	2.282
Desired Nutritional Health Status		
a. Interpretation of clinical nutritional biochemical tests	2.083	2.146
b. Recommended Dietary Allowances		2.154
Scientific Methodology for Nutrition and Dietetics		
a. Anthropometric Measurement		2.108
c. Nutritional Reference Values		2.468
Domain of Human Condition		
Pre-Clinical Sciences		
b. Biochemistry	1.923	2.763
c. Biology	1.992	2.603
d. Physiology	2.198	2.580
e. Psychology	2.092	1.649

Knowledge Topics Considered be Shared Knowledge Derived from Scores of Relevance to Clinical Dietetics (≥ 2.5) and Comparative Relevance to Other Health Professionals and Nutritional Scientists (> 1.5)

Knowledge Topic and Domain	Comparative Relevance Scores Knowledge Shared With	
	OHP*	NS**
Knowledge Concerning Individuals' Departures From Normal Human Functioning		
a. Changes in tissue appearance related to nutritional deficiency........	1.985	2.379
b. Individual's food allergies/sensitivities.................................	1.727	1.577
c. Signs/symptoms/potential for disease development related to nutritional status/function	1.870	2.138
b. Interaction of lifestyle status, health status, with nutritional status ..	1.661	1.646
Nutritional Status		
c. Prognostic Nutrition Indices...		1.845
d. Potential or Presence of Nutrient/Food/Drug Interactions	1.984	2.015
Domain of Practitioner Actions/Attitudes		
Assessment of Individuals		
c. Health status (medical dx, Rx, health/dental problems, mobility, vision, biochemical)	2.177	1.554
d. Mental status (readiness to change, educational, level, cognitive functioning)...	1.915	
e. Nutritional status (dietetic-specific physical examination, anthropometric)...		1.585
Nutritional Diagnosis		
a. Diagnostic thought processes	1.819	1.692
b. Evaluation for etiologies...	1.612	1.706
e. Intellectual Multi-skilling: (In-depth, science-based clinical practice, critical/global thinking, conceptualizing)........................	1.854	2.023
Participation in Clinical Dietetics' Research		
a. As collaborator...	1.508	2.546

Knowledge Topics Considered be Shared Knowledge Derived from Scores of Relevance to Clinical Dietetics (≥ 2.5) and Comparative Relevance to Other Health Professionals and Nutritional Scientists (> 1.5)

Knowledge Topic and Domain	Comparative Relevance Scores Knowledge Shared With	
	OHP*	NS**
Philosophy of Clinical Dietetics		
a. Practitioner's personal professional philosophy		1.515
Relationships		
b. Networks	1.779	1.756
c. Interdependent Colleagues	1.832	1.731
Domain of Practitioner Environment		
Political/Social Milieu		
b. Public policy/law	1.739	1.736
d. Funding for research in Clinical Dietetics		2.515
Available Tools and Technology		
a. Tools for physical nutritional assessment of individuals		1.891
b. Tools for assessment of nutrition-related behaviors of individuals	1.701	
c. Tools for attitude assessment of individuals	1.626	1.632
d. Computer programs re: nutrition and/or diet		1.960
e. Electronic communication capability	1.771	1.976
Domain of Client Actions/Attitudes		
Acceptance of Responsibility for Self Care		
a. Ability for nutritional self care	1.638	
c. World view re: health and nutrition		1.908
Influences on Food Intake		
b. Emotional status	1.856	

Knowledge Topics Considered be Shared Knowledge Derived from Scores of Relevance to Clinical Dietetics (≥ 2.5) and Comparative Relevance to Other Health Professionals and Nutritional Scientists (> 1.5)

	Comparative Relevance Scores Knowledge Shared With	
Knowledge Topic and Domain	OHP*	NS**
Choices Made		
a. Lifestyle choices	1.756	
Domain of Domain of Client Environment		
Food Resources		
b. Route of food delivery (enteral, parenteral)	1.840	
Personal Social Resources		
a. Adequate finances for food..	1.672	
b. Transportation to acquire food..	1.591	
d. Care-giver who is responsible and knowledgeable re: nutrition needs...	1.746	
Environmental Threats		
a. Food-borne environmental toxins	1.734	2.085
b. Misleading nutritional claims..		1.674
Domain of Nutraceuticals		
Knowledge Concerning Nutraceuticals		
a. Composition of Nutraceuticals...		2.446
Appropriate Use of Nutraceuticals		
a. Functions of Nutraceuticals...	1.527	2.423
b. Effects of Nutraceuticals...	1.634	2.409

*OHP=Other Health Professionals   **NS=Nutritional Scientists

Other Health Professionals

Respondents were asked to indicate which other health professionals they were considering when indicating Comparative Relevance. Four professions were listed (Nurses, Pharmacists, Physicians and Psychologists) and a space provided to list any other health professionals respondents considered. Multiple answers were allowed.

An average of eighty-five percent of respondents indicated they were considering nurses, forty-four percent considered pharmacists, seventy-five percent considered physicians, twenty-percent considered psychologists and twenty percent considered other categories of health professionals when scoring comparative relevance of knowledge topics.

Categories of Other Health Professionals added by respondents included the following:

1. Social Workers were named by twenty-four respondents. Twenty named Social Workers under the domain of Client Environment.

2. Physical Therapists were named by sixteen respondents.

3. Occupational Therapists were named by thirteen respondents.

4. Speech Therapists were named by twelve respondents.

Additional types of professionals named by four or less respondents included: activity therapist, behaviorist, biochemist, certified diabetic educator, dialysis technician, non-clinical dietitian, exercise physiologist, food scientist, health educator, specific nurse categories of NP, FNP, LNP, nurse discharge planner, and staff nurses, nutritionist, physician assistant, physiologist, rehabilitation therapist, researcher, respiratory therapist and therapist.

Equally Relevant Knowledge Topics

Knowledge Topics perceived as Relevant to Clinical Dietetics (relevance score of 2.5 or greater) and equally relevant (comparative relevance score of 2.0-2.5) to Other Health professionals include:

> Aging
> Growth patterns
> Physiology

Psychology

Interpretation of clinical nutritional biochemical tests.

Knowledge topics perceived as Relevant to Clinical Dietetics (relevance score of 2.5 or greater) and equally relevant (comparative relevance score of 2.0-2.5) to Nutritional Scientists include:

Aging

Growth patterns

Signs/symptoms of disease development

Potential for nutrient/drug interaction

Intellectual multi-skilling

Food borne environmental toxins

Interpretation of clinical nutritional biochemical tests, RDA

Knowledge topics perceived as Relevant to Clinical Dietetics (relevance score of 2.5 or greater) and equally relevant (comparative relevance score of 2.0-2.5) to Nutritional Scientists include, continued:

Anthropometric measurement and

Composition, Function and Effects of nutraceuticals.

Knowledge Topics More Relevant to Nutritional Scientists

Knowledge topics perceived to be relevant to clinical dietetics, (relevance score of 2.5 or greater) but of greater relevance to nutritional scientists (comparative relevance score of 2.5 or greater) include:

Nutrient disposition

Nutrient function

Integrated metabolism

Biochemistry

Biology, physiology

Tissue appearance in nutritional deficiency

Participating in research as a collaborator.

No knowledge topics were perceived to be relevant to clinical dietetics and more relevant to other health professionals.

Summary regarding shared knowledge topics

77 of 94 knowledge topics were considered relevant to clinical dietetics (82%).

27 of 94 knowledge topics were perceived as unique to clinical dietetics (29%).

49 of 94 knowledge topics were perceived as relevant and shared (52%).

> 26 of these 49 were shared with both other health professionals and nutrition scientists (28% of total knowledge topics).
>
> 16 of these 49 were shared with nutritional scientists (17% of total knowledge topics).
>
> 7 of these 49 were shared with other health professionals (7% of total knowledge topics).

Comparison of Relevance by Domains of the Metaparadigm

Comparisons between clinical dietetics, other health professionals and nutritional scientists can also be made on the basis of mean scores of relevance and comparative relevance for individual domains as well as individual knowledge topics. Respondents scored all seven domains as having mean relevance scores greater than 2.5 for Clinical Dietetics. Comparative relevance to Other Health Professionals scores indicated equal relevance for the domains of Reference Person, Human Condition, Client Actions/Attitudes and Client Environment. Of less comparative relevance to Other Health Professionals were the domains of Practitioner Actions/Attitudes, Practitioner Environment and Nutraceuticals. Nutritional Scientists were scored as having equal comparative relevance in the domains of Reference Person, Human Condition, Practitioner Environment and Nutraceuticals and less comparative relevance in the domains of Practitioner Actions/Attitudes, Client Actions/Attitudes and Client Environment.

Results and Discussion

Possible sources of bias in the study

The final sample of N=136 represented a response rate of twenty-seven percent. This response rate is not adequate for generalization to all clinical dietetic practitioners in the American Dietetic Association.

The sample contained no respondents who had been in practice for less than five years which may have led to a small number of those entering the profession through the AP4 route of entry which is the newest route of entry. These more recently educated respondents could possibly perceive knowledge relevant to the profession differently than more seasoned practitioners, thus introducing bias into the results. It may be especially important in the future to determine the perceptions of additional AP4 graduates, since scores of relevance of the two respondents for the domains of Practitioner Actions/Attitudes and Nutraceuticals were strikingly different from those entering the profession through the experience and internship routes.

Eighty-one percent of respondents had been practicing clinical dietetics longer than ten years, almost half the sample had been in practice longer than twenty years. This sample was seasoned professionals with a long term perspective of the profession and the milieu in which clinical dietetics has existed. It is possible that the initial letter requesting commitment to participate in the survey appealed to experienced practitioners, but not new practitioners.

The northwest and southwest regions of the United States were under-represented in the final sample with these two areas providing five percent of the responses. The distribution of respondents' location did not closely resemble the initial sample, which may have introduced some bias. Dietitians in less populated states may travel further to practice, may practice alone more often and may have to deal with more diverse populations, which may influence their perceptions of practice.

Sixty-one percent of respondents belonged to four Dietetic Practice Groups representing General Clinical Practice, Health Care Facility, Gerontological and Public Health practices. Relatively strong representation from the General Clinical Dietetics DPG (20% of respondents) was appropriate since this study addresses the Metaparadigm of Clinical Dietetics. The hospital practice environment was reported by forty-six percent of respondents with an additional fifteen percent practicing in long-term care facilities. Out-patient and private practice was the reported environment of thirty-three percent of practitioners. This seems an intuitively fair representation of current practice environments, with projected growth environments being Gerontology and Public Health. A sample containing more specialty clinical environments and the potential future growth area of managed care environments would create a broader representation of clinical dietetics for the future.

Respondents were highly educated in science: seventy-eight percent of respondents had degrees in science: B.S., M.S., or D.Sc. Forty-five percent has obtained a Master's Degree, 5.5% has doctorate degrees. This orientation toward science supports science-based practice, professional advancement and building a profession-specific scientific body of knowledge.

Survey Instrument

Cronbach alpha reliability scores of ≥ 0.70 for a new instrument combined with 1) respondent and expert panel perceptions of relevance, 2) the ability of respondents to place their work in the proposed domains and 3) low number of suggestions for additional knowledge topics allow a conclusion of construct validity for this sample of practitioners for the construct Metaparadigm of Clinical Dietetics. Results indicate a high degree of content validity.

Relevance of Metaparadigm and Knowledge Topics

Metaparadigm

Ninety-five percent of respondents perceived the proposed Metaparadigm of Clinical Dietetics as encompassing the professional body of knowledge. Sixty-five percent placed their own work in four or more of the proposed domains of the Metaparadigm of Clinical Dietetics. Six of the seven domains were selected by over fifty-percent of respondents as areas in which they placed their own work. This indicates that using the direct questioning method, the domains and knowledge topics which characterized them were perceived as inclusive and applicable by clinical dietetic practitioners, validating the proposed Metaparadigm of Clinical Dietetics.

Relevance of Knowledge Topics

Eighty-two percent of the knowledge topics were perceived as relevant to clinical dietetics (relevance score of 2.5 or greater). This produces a skewed distribution but also indicates that the selection of knowledge topics was not trivial and that the instrument has power of discrimination.

The knowledge topic with the highest relevance score and least variance was "Nutrition Counseling". Only one knowledge topic was perceived as Somewhat Irrelevant: "Technical Multi-skilling", characterized by taking blood pressures and drawing blood. This indicates practitioners see their most relevant role as nutrition counselor rather than nutrition educator and see some technical skills as domains of other professions. A newly developing role as the provider of medical nutrition therapy, (nutrition therapist) should be included in any future validation of the Metaparadigm of Clinical Dietetics.

Twenty-seven of the ninety-four knowledge topics (29%) from five domains were perceived to be unique to clinical dietetics, that is, to have relevance to clinical dietetics and less comparative relevance to other health professionals and nutritional scientists. The two domains not containing any knowledge topics perceived as unique were Reference Person and Nutraceuticals. This is not surprising since standards of reference for "normal" and "health" are appropriately shared, though the terms are not defined precisely across disciplines. Newly developing areas within the domain of Nutraceuticals are not yet firmly associated with any profession or group. Forty-two percent of respondents placed their work in this domain. Including food in this category may not coincide with how practitioners normally think of nutraceuticals, even though this was included in the definition provided. Differentiation between unique and shared knowledge topics appear to be logical and serve to further demonstrate the discriminating power or construct validity of the survey instrument. Results of this study support the original clinical dietetic research typology of five domains by Kight.

Unique Knowledge Topics

A synopsis of topics considered unique to clinical dietetics includes 1) use of nutrition diagnostic codes, 2) thorough in-depth assessment of clients, 3) use of dietetic interventions, 4) documentation of interventions and outcomes, 5) management of clinical dietetic services, 6) acting from a sense of professional philosophy and responsibility, 7) credentialing, 8) meeting standards of care, and 9) use of nutritional education materials. The unique aspects of clinical dietetics that is demonstrated indicates that Practitioner Actions/Attitudes and how they are enacted relative to the Human Condition is where professional uniqueness is perceived.

Notably missing from this list is theory-building in clinical dietetics or acting as principal investigator (perceived as Somewhat Relevant to clinical dietetics). Profession-specific theories and research are two tools for building a profession-specific body of knowledge. Respondents perceived theory-building in clinical dietetics as somewhat relevant (relevance score of 2.3) and equally as relevant to nutritional scientists as to clinical dietetics (comparative relevance score of 2.1). This indicates a need for inclusion of clinical dietetic meta-theory and middle-range theory as well as the transformation of nutritional science micro-theories into clinical dietetic practice micro-theories in clinical dietetic preparatory and continuing education. The identification, development, and application of theory to practice as well as explicit processing of the value derived from theory-building are potential components of such education.

Respondents perceived the role of research collaborator as relevant, the role of principal investigator as somewhat relevant. The role of principal investigator is perceived as more important to nutritional scientists. The involvement in research, especially present research, is predictive of a difference in mean relevance scores for the domains of Reference Person and Human Condition. Practitioners may define research as laboratory-oriented nutritional research. This indicates clinical dietetic practitioners need knowledge and support (financial, educational and mentoring support) to develop interest and skill in clinically-oriented research. To develop clinical dietetic knowledge, practitioners need to see their practice environment as the research laboratory. The amount of knowledge perceived as shared with nutritional scientists was somewhat greater than that shared with other health professionals. Increase of clinical dietetic research affiliation with other health professionals in the application of nutritional science in the clinical environment seems the next step in development of professional knowledge. This step has potential for being personally rewarding, professionally enhancing and socially valuable.

Shared Knowledge Topics

Twenty-eight percent of the knowledge topics were shared with both nutritional scientists and other health professionals. Seventeen percent of knowledge topics were shared with only nutritional scientists, seven percent with only other health professionals, indicating a perception of greater alignment with nutritional sciences than with health care professionals at this time.

Respondents considered a variety of other health professionals when scoring knowledge topics for comparative relevance. Eighty-five percent scored comparative relevance while thinking of nurses, seventy-five percent considered physicians, forty-four percent considered pharmacists, twenty percent considered psychologists. Eighteen percent named social workers, twelve percent named physical therapists, ten percent named occupational therapists and speech therapists were named by nine percent. This illustrates the diverse milieu in which clinical dietetic practitioners find themselves and with whom they are making role comparisons.

These indicators of relevance, uniqueness, shared knowledge and numbers of other health professionals considered by practitioners leads to the conclusion that boundaries (and therefore, potentially, roles) are indeed ambiguous. Distinguishing the extent and areas of shared knowledge gives the potential for educators to plan cooperative course work with other health professionals and nutritional scientists as recommended by the Pew Commission (Pew Health Professions Commission, Health Professions Education, 1993) (Pew Health Professions Commission. Healthy American: Practitioners for 2005, 1991) when projecting educational needs for health professions. Joint course work would decrease duplicate efforts, setting the stage for formation of collegial relationships and cooperative research. Increased recognition by other health professionals of the clinical dietitian's unique contribution to health care might also result.

Factor Analysis

Factor analysis is a statistical method that indicates a quality of "alikeness" of individual items in groups that are established by statistical correlation (Kline, 1994). The level of correlation between individual items accepted for inclusion in a group is established by the investigator.

Perfect correlation between two items is 1.0. For example, 100% of the time B occurs in the data, J also occurs in the data: they are perfectly correlated. An example of a high level of correlation is r = 0.7 and above. An example of a low level of correlation is r = 0.2.

Once groups of items/answers/responses are established by correlation, the investigator looks at the contents (items in each group) to ascertain what is the common theme or factor that could be used to name the group based on what characteristic the items in each group share.

A simple factor analysis example is included to illustrate the principle of this statistical method.

The items in the metaparadigm validation study were the scores of relevance to clinical dietetics of the 94 knowledge topics and scores of comparative relevance to nutritional scientists and other health professionals.

Factor analysis of perceptions of 136 respondents' regarding 94 knowledge topics

The factor analysis of the scores from the metaparadigm survey can be considered as yielding preliminary conclusions: an emerging theory of how clinical dietetic knowledge can be perceived related to the Eight World Hypotheses. A database using responses from a greater number of practitioners would indicate whether the content factors (knowledge topics) would remain consistent or whether they would shift into a different pattern that might require different labels. A minimum sample size of 235 respondents is preferred for a stable factor structure for this number of items (94). The sample size for this study was 136. Knowledge topics with loadings indicating a strong positive relationship (r = ≥ 0.50) with the factor were used in this interpretation. The seven factors in this analysis, accounted for approximately forty percent of the variance. This means that 60% of the variance in the factor analysis was produced by unidentified influences. Explaining more of the variance by using a larger sample size is needed.

<u>Example illustrating factor analysis:</u>

A mixed group of fruits and vegetables exists: (the items)

Onions, bananas, grapes, oranges, celery, peas, eggplant, tangerines, jicama, corn, tomatoes, beets, apples, cucumbers, lemons

A factor analysis has created the following groups; the correlation of the item to the group is part of the factor analysis:

I	II	III	IV	V	VI
Bananas	Tomato	Oranges	Eggplant	Onions	Peas
Cucumbers	Apples	Lemons	Grapes	Jicama	Corn
Celery	Beets	Tangerines			

Factor I could be labeled by shape: long and narrow fruits and vegetables; bananas and cucumbers would likely be highly correlated, celery would be less related
        A Factor Label: Long and Narrow Fruits and Vegetables

Factor II could be labeled by color: red: tomato and beets would be highly correlated because the color goes through the whole vegetable
        A Factor Label: Red Fruits and Vegetables

Factor III could be label by flavor: all are fruits that have citrus flavor
        A Factor label: Citrus Fruits

Factor IV could be labeled by color or by shape: purple-ness or round-ness
        A Factor Label: Round Purple Fruits and Vegetables

Factor V could be labeled by color: white-ness
        A Factor Label: White Vegetables

Factor VI could be labeled by shape, size, how they grow (numerous items in a pod or on a cob) or most frequently served in a certain type restaurants in the United States
        A Factor Label: Vegetables Frequently Served in Diners in the US

Other potential items might fit into two groups: blueberrries could fit in the purple factor or the small and round factor.

If an investigator observed other similarities that were more significant to the hypothesis, factors could be labeled according to other qualities. This could include cost in the supermarket, what season they become ripe, where they are grown, what kind of plant they grow on, sweetness of flavor, etc. The purpose of the factor analysis and the interpretation of the investigator will determine the end results.

Eight World Hypotheses

To describe ways of understanding the world and the phenomena we experience and observe, four World Hypotheses were proposed by Stephen Pepper in 1942 (Pepper, 1942) These four world hypotheses were 1) Formistic, 2) Mechanistic, 3) Contextual, 4) Organismic. An additional four World Hypotheses describing new ways of understanding the world through the eyes of science since 1942 were proposed by Gary Schwartz and Linda Russek in 1997 (Schwartz & Russek, 1997). The additional four of the eight World Hypotheses are described as 5) Implicit Process, 6) Circular Causality, 7) Creative Unfolding and 8) Integrative Diversity. Each level is included and integrated into the next higher level of organization; each level increases in complexity of knowledge. This organization can apply to perceptions of how the physical world is organized or can be applied to non-physical phenomena such as thought processes.

A Brief Description of Each World Hypothesis:

World Hypothesis I—Formistic: The Formistic hypothesis sees structures and functions existing as separate categories.

World Hypotheses II—Mechanistic: The Mechanistic hypotheses assume that all effects have causes that precede them. These may also include processes involving whole categories.

World Hypothesis III—Contextual : The Contextual hypothesis assumes that all structures and functions exist in context and are relative, not absolute.

World Hypothesis IV—Organismic: The Organismic hypothesis sees the world as systems of interacting structures and functions, parts interacting to become a larger whole system. The system, or "whole", is seen to be greater than the simple sum of the parts which make it up.

World Hypothesis V—Implicit Process: The Implicit Process hypothesis states that all systems involve invisible processes that interact over time. These invisible processes may be information, energy and/or matter.

World Hypothesis VI—Circular Causality: The Circular Causality hypothesis states that the processes that are invisible and interact over time <u>change</u> because of this interaction through time. Feedback loops which modulate or regulate a reaction, a behavior, or storage of information are examples of this hypothesis.

World Hypothesis VII—Creative Unfolding: The Creative Unfolding hypothesis states that systems create by intent, planning, order and growth.

World Hypothesis VIII—Integrative Diversity: The Integrative Diversity hypothesis states all phenomena in nature reflect complex interconnected, integrated orders of diverse processes, an example being the developing alternative medicine sciences. (Schwartz & Russek, 1997)

Levels of Knowledge Complexity in the Metaparadigm of Clinical Dietetics

Knowledge complexity increases from the center outward as each new level incorporates the components of less complex levels.

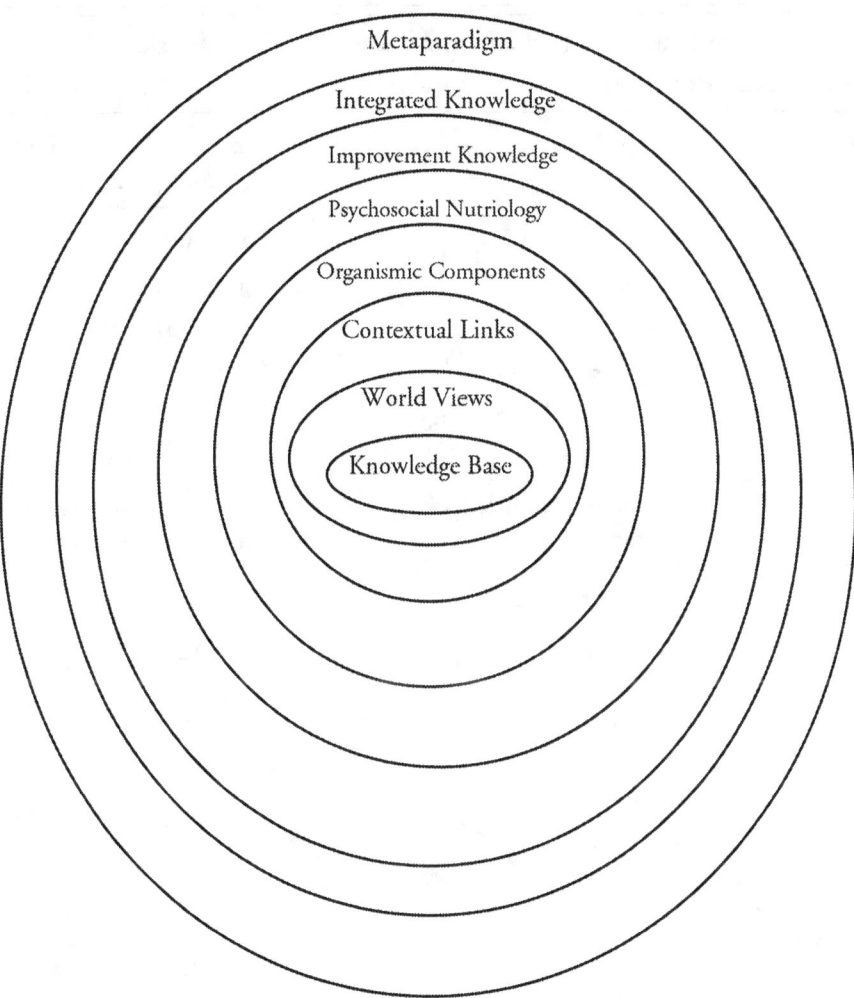

Relating Schwartz and Russek World Hypotheses Wheel to the Body of Knowledge of Clinical Dietetics as Revealed by Factor Analysis.

Induced Clinical Dietetic Factor	Factor Label	World Hypotheses Link (Increasing Levels of Complexity)
VI	Knowledge Base	Formistic
II	World Views	Mechanistic
VII	Contextual Links	Contextual
I	Organismic Components	Organismic
III	Psychosocial Nutriology	Implicit Process
V	Improvement Knowledge	Circular Causality
IV	Integrated Knowledge Metaparadigm	Creative Unfolding Integrated Diversity

7. Integrated Knowledge (Creative Unfolding)

1. Knowledge Base (Formistic)

6. Improvement Knowledge (Circular Causality)

8. Integrated Diversity (Metaparadigm)

2. World Views (Mechanistic)

5. Psychosocial Nutriology (Implicit Process)

3. Contextual Links (Contextual)

4. Organismic Components (Organismic)

Metascopes

The metascope illustrates which domains of the Metaparadigm of Clinical Dietetics contributed knowledge topics to which World Hypotheses Labels (WHL) as determined by a factor analysis of the ninety-four knowledge topics and their correlations (Clinical Dietetic Factors: CDF).

This figure of the Metascope illustrates links that the Clinical Dietetic Factor (CDF) labeled "Knowledge Base" contained knowledge topics from the Metaparadigm of Clinical Dietetics domains of Reference Person and Human Condition.

The World Hypothesis Label 7 (WHL 7) (CDF label IV)—"Integrated Knowledge"—contains knowledge topics from the domains of Reference Person, Human Condition and Practitioner Actions/Attitudes.

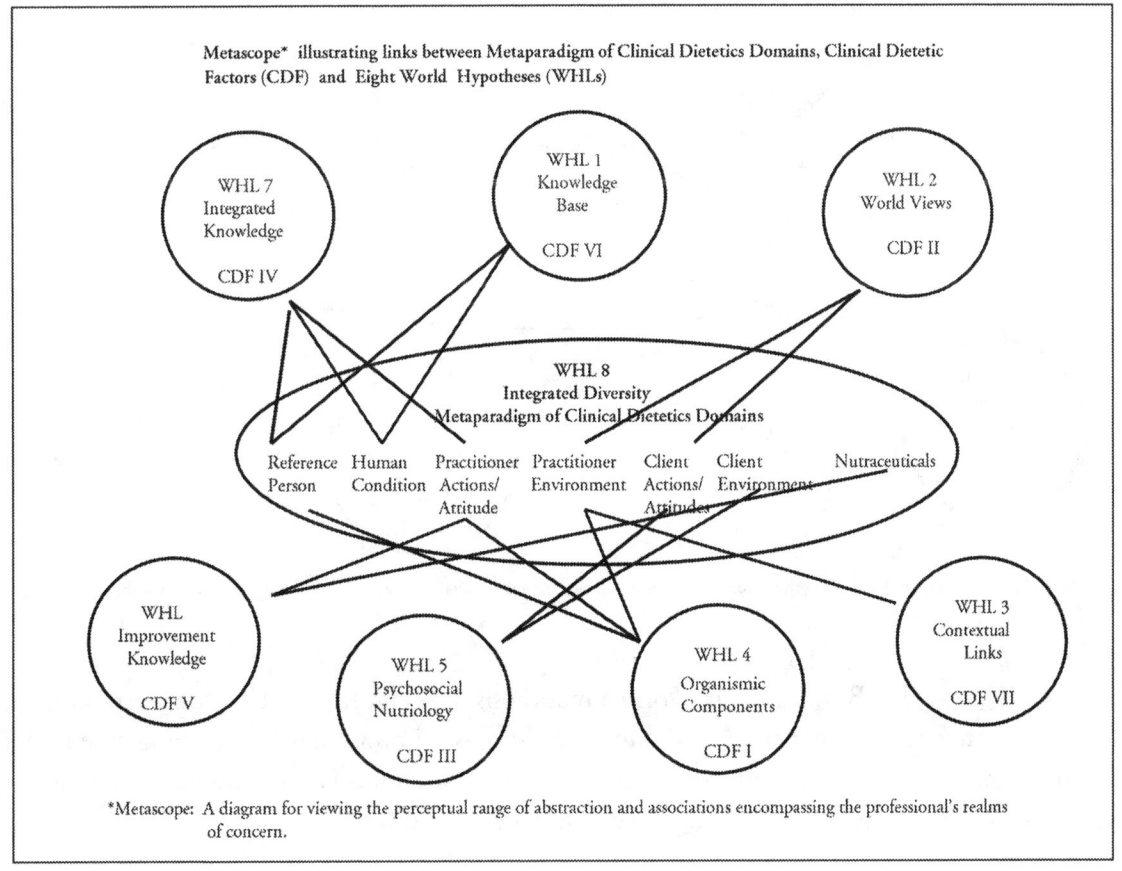

Metascope* illustrating links between Metaparadigm of Clinical Dietetics Domains, Clinical Dietetic Factors (CDF) and Eight World Hypotheses (WHLs)

*Metascope: A diagram for viewing the perceptual range of abstraction and associations encompassing the professional's realms of concern.

This metascope, illustrating links with topics perceived as unique, illustrates that the domains perceived as containing the knowledge topics unique to clinical dietetics contribute to the Clinical Dietetic Factors of: 1) Knowledge Base, 2) Organismic Components, 3) Improvement Knowledge and 4) Integrated Knowledge. This can be interpreted as supporting the theory that knowledge topics perceived by practitioners as unique to Clinical Dietetics parallel the Eight World Hypotheses as follows:

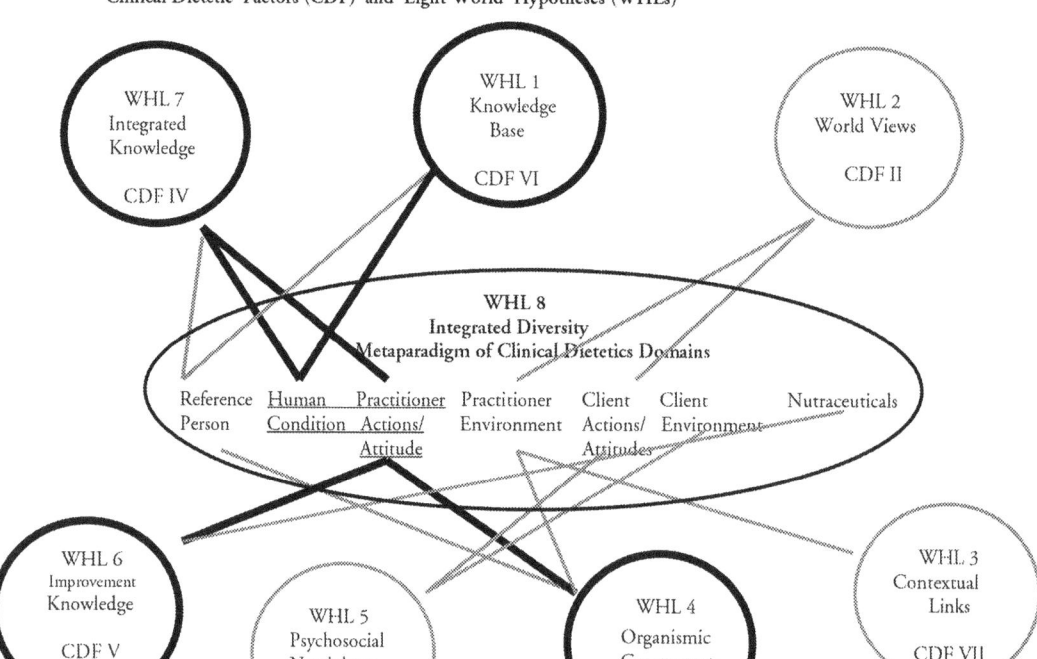

Metascope* illustrating links between Metaparadigm of Clinical Dietetics Domains Perceived as Unique , Clinical Dietetic Factors (CDF) and Eight World Hypotheses (WHLs)

*Metascope: A diagram for viewing the perceptual range of abstraction and associations encompassing the professional's realms of concern.

CD Factor <u>Knowledge Base</u> parallels World Hypothesis 1—The first level of understanding the world in by categories of structure and function, ie, basic knowledge of assigning phenomena into categories, naming them and describing how they function. The phenomena of Nutrition Diagnostic Codes are categories describing the Human Condition.

CD Factor <u>Organismic Components</u> parallels World Hypothesis IV—The organismic view is a more complex way of understanding the world as systems of interacting structures and functions, creating a system that is greater than the simple sum of its parts. The human body systems and nutritional interactions can be understood in this way.

CD Factor <u>Improvement Knowledge</u> parallels World Hypothesis V—The Circular Causality Hypothesis understands the world to contain processes that are invisible and that interact over time, ant this interaction over time causes change. Feed back loops that change and control/maintain homeostasis is an example. Nutrients can have this physical function (glucose/insulin). Invisible processes such as empathy, education and Practitioner/Client interaction can be seen as an example of improvement knowledge which can change behavior as well as biochemistry.

CD Factor <u>Integrated Knowledge</u> parallels World Hypothesis 7—The Creative Unfolding hypothesis states that systems create by intent, planning, order and growth. The system of the Practitioner and Client create a change in behavior, attitude, and/or health by meeting with the intention for planning change and a growth toward health. The interactions between the Practitioner and Client unfold and change as treatment progresses, initial goals are met and altered to plan for a continuum of progress.

Metascope Example Using Biesecker's Theory

Biesecker's theory is a description of inner processes (Attitude) in Practitioner development of the emerging quality of expertise.

Metascope of theory of Ronna Biesecker PhD,RD:

Links between

Domains of Metaparadigm of Clinical Dietetics and

expanded Labels for Clinical Dietetic Factors which link domains with the

Eight World Hypotheses

This metascope illustrates that a theory concerning Practitioner's Attitudes (the quality of expertise) has potential for building improvement knowledge, organismic knowledge building integrated knowledge.

Building Improvement Knowledge may be thought of as the research process of defining how to become more of an expert and how to recognize or measure expertise.

Building Organismic Knowledge may be thought of as research which would discover or describe the cause and effects that are present in creating or using expertise.

Building Integrated Knowledge may demonstrate how increasing expertise in the profession as a whole, or as individual practitioners, can be integrated into clinical practice, perhaps improving clinical health outcomes or expanding the role of the clinical dietitian.

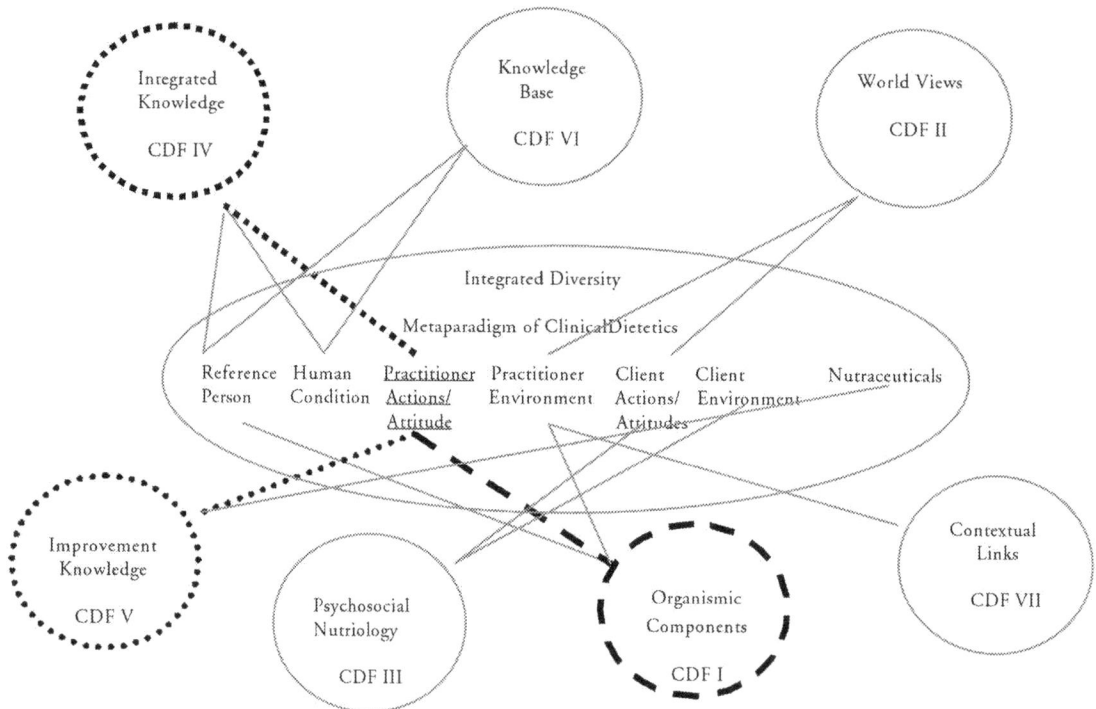

II. Summary: Developing Knowledge with Theory

Developing the body of knowledge "guided by theory rather than random efforts to relate things to one another ... reduces the trial-and-error effort ... People who explore theories are at the vanguard of each field of science ... It is just as important to develop theories regarding attributes as it is to develop methods of analysis" (Nunnally & Bernstein, 1994). Leaders in the dietetic profession recently commented concerning the need for research in clinical dietetics for supporting development of the profession as well as individual practitioners (Dwyer, 1997) (Fitz, 1997). The Metaparadigm of Clinical Dietetics, linked with the Eight World Hypotheses, has the potential for serving as an organizing influence for creation and explication of profession-specific theories. The profession of nursing has demonstrated the influence of a professional metaparadigm on practice and research (Fawcett, 1995). Blegan and Tripp-Reimer (1997) comment that the four concepts in the Metaparadigm of Nursing are central to other disciplines as well, ... that they have "provided a rationale and mechanism for the discipline to move beyond the conceptual models" Clinical Dietetics' route to knowledge development will have some similarities to and differences from the route of nursing knowledge building.

Clinical dietetic practitioners have potential for transforming nutritional science knowledge into clinical nutrition knowledge, for developing new knowledge in collaboration with nutritional scientists and acting as principal investigator or colleagues of other health professionals in clinical research. Advanced level practice can be expected to facilitate developing this potential. Clinical nutrition has been recently differentiated from nutritional epidemiology, with the two seen as being on a continuum from intake to disease (Forman & Halsted, 1998) (see Glossary). Clinical dietetic practitioners can appropriately be involved in the study and treatment of both the large populations of epidemiology and the smaller numbers of direct patient care.

For creating science-based, profession-specific theory around areas of uniqueness, the Metaparadigm of Clinical Dietetics and the Eight World Hypotheses are recommended for building the body of knowledge of clinical dietetics—clinical knowledge, educational knowledge and research knowledge. In the process, practitioners can make contributions to solutions needed for effective health care while realizing individual professional satisfaction.

SECTION B

Applications of the Metaparadigm of Clinical Dietetics

The domains of the Metaparadigm of Clinical Dietetics are guides for thinking about knowledge and clinical dietetics practice. A guide implies the "how" of thinking, the "structure" of thinking. The "phenomena of concern" of a profession are the situations, topics, areas, problems and general activities of the individuals practicing the profession. The knowledge topics, or "characterizations", of the seven domains provide specific topics within each domain which are of concern to clinical dietetics practitioners.

The following examples of applications of the Metaparadigm of Clinical Dietetics are hypothetical illustrations of how the domains may structure thought in practice, education and research. They are derived from the author's experience in clinical dietetic practice and graduate education. These applications are intended to demonstrate and suggest how the guide of the Metaparadigm of Clinical Dietetics may be incorporated into the practice, education and research of clinical dietetics. Any of these ideas may be extended, modified, applied and tested by those interested in using the metaparadigm to structure and theoretically support their thinking.

III. Dietetic Practice

A. Using Metaparadigm Domains in Creating Clinical Dietetic Treatment Matrices: If … Then … Then … Guides to Practice

Using If … Then … Then … statements are a process of human thinking using logic. This process links observations of a situation or condition to drawing conclusions to taking action.

$$\text{Observation} \longrightarrow \text{Conclusion} \longrightarrow \text{Action}$$
$$\textbf{If} \dots\dots\dots\dots\dots \textbf{Then} \dots\dots\dots\dots \textbf{Then}$$

A health care professional uses If … then … then … logic when making a diagnosis or decision regarding the condition of the client. A clinical dietitian reaches a nutritional diagnosis; a physician reaches a medical diagnosis; a nurse reaches a nursing diagnosis.

Combined with an X and a Y axis format, If … then … then … thinking can be portrayed as decision-making treatment matrix.

$$\text{Then} \dots \quad | \qquad Y$$

$$\overline{\qquad\qquad}$$
$$\text{If} \dots \quad X$$

Two domains of the Metaparadigm of Clinical Dietetics can be used to name the X and Y axis of the treatment matrix. If … a specified Human Condition is observed or presents for nutritional interventions, Then … Practitioner Actions (best practice) can be portrayed.

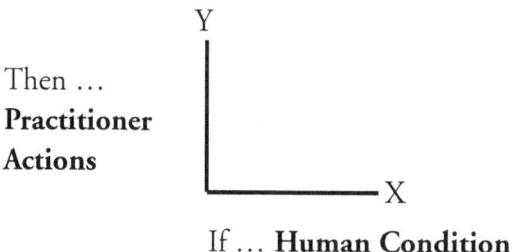

If ... **Human Condition**

A practice matrix for any health problem treated by a clinical dietitian has the potential for portrayal in this format. Agreement among practitioners concerning which actions are most appropriate for a given condition can provide a quick, easy-to-use guide for nutritional care protocols.

As the Human Conditions progress from left to right along the X axis, they become less life-threatening. The Practitioner Actions can be thought of as progressing through time as they progress along the Y axis.

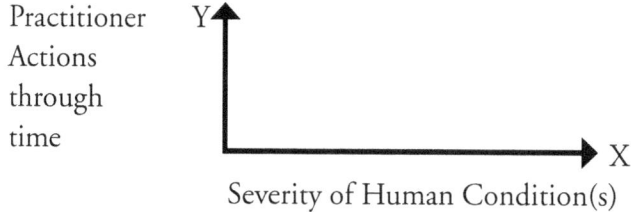

Severity of Human Condition(s)

Observation and identification of the Human Condition(s) can be conceived of as (going from left to right): 1) most life-threatening at the intersection of X and Y (for example, life-threatening levels of K) to 2) those having longer term, but not immediately life-threatening impact on the patient's current health problems (for example dietary intervention for chronic renal failure), progressing to 3) disease prevention (for example weight loss for individuals with a family history of diabetes), and then to 4) supporting optimum health (for example appropriate weight gain during pregnancy) at the most distant point on X.

In reality, several conditions may occur at the same time, may come and go in different combinations and/or at different times. Considering the severity of symptoms being treated at a given time can be helpful in setting priorities for interventions.

The Practitioner's Action can be conceived of as moving from bottom to the top of the Y axis, 1) the concrete being at the intersection of the X and Y axis (such as performing a physical examination), up the Y axis to 2) less concrete (such as calculating energy needs). A Practitioner might then proceed to 3) recommend/order a specific treatment procedure or modality (for example specify a given number of calories or grams of protein, rate of tube feeding flow, oz. of fluid, omission of a food to which a suspected intolerance exists, content and frequency of meals, etc.) Continuing up the Y axis, actions may proceed to 4) negotiating with the patient concerning behavioral changes recommended (which behaviors, how to accomplish, at what pace, measures of success). Further up the Y axis the Practitioner Actions might focus on 5) therapeutic use of self (forming and maintaining a relationship with the patient) aimed at providing support in obtaining behavioral or educational goals. Therapeutic use of self might also be placed at the bottom of the Y axis, as the starting point, depending on the practice environment, the nutritional care provided and the perspective of the individual practitioner.

Also on the Y axis, though less concrete, Practitioner Actions on behalf of the patient might include 6) making a referral to an RD specialist, a psychologist, an MD, etc., or arranging follow-up office visits.

Progression up the Y axis in Practitioner Actions will be influenced by the practice environment. Actions in hospital critical care will vary considerably from actions in an out-patient setting treating eating disorders, for example. Actions will also be influenced by the pattern of Human Conditions the treatment matrix addresses. A renal nutritional care matrix will vary considerably from the matrix describing the post-surgical progression from liquids to a regular diet.

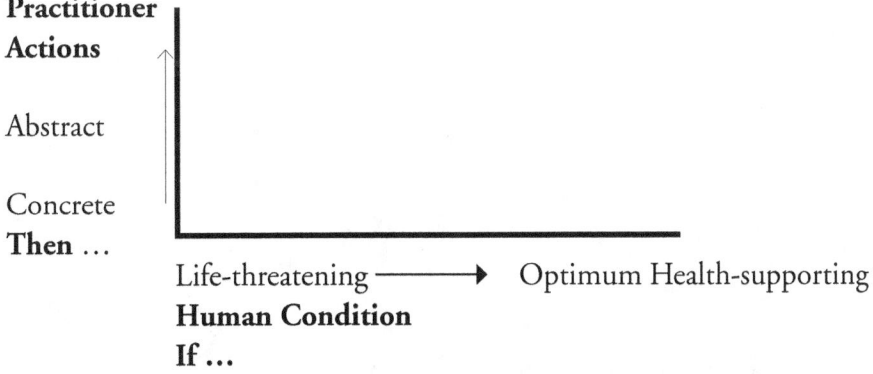

Treatment matrices may include criteria for diagnosis and successful outcomes on the matrix itself or be detailed on a separate page for sake of providing necessary details.

An example of one type of treatment matrix follows:

Then …	Hx: Symptoms, Diet, Family, Allergies, Medical	Nutritional Diagnosis: Alteration in bowel function 2° lactase insufficiency	Tx/care plan Recommendations: Omission of dairy foods; Education	Assess Outcome; Follow-up appt:1 week

Practitioner Actions

If … Human Condition : Frequent diarrhea

A second example of a detailed treatment matrix, for Eating Disorders follows. Pages 1 & 2 are to be read as a continuous unit. Pages 3 & 4 are also one continuous unit and extend the X axis from pages 1 & 2. Pages 3 & 4 are the extension through time of pages 1 & 2. The Treatment Protocol that follows the treatment matrix provides the specific indicators for defining a problem and desired outcomes of the intervention.

Dietetic-Specific Eating Disorder Protocol—Bulimia, Anorexia Nervosa, Compulsive Overeating—Master Matrix

Y Axis, Then ↓ — Practitioner Actions/Attitudes Interventions							
Schedule follow-up	X	X		X	X	X	
Biochemical							X
Refer to Psych If not in Tx	X	X					
Refer to MD-PCP	X			X			
Negotiated Goals: Client:RD							
Wt: Long term							
Rate of change	X	X		X			
Calorie intake	X	X	X	X			
Protein Intake/form	X	X	X				
Carbohydrate intake/form	X	X		X			
Fat intake/form	X	X		X			
Fiber intake	X	X		X			
Fluid intake	X	X		X			
Exercise level	X			X	X		
Vitamin/Mineral supplements					X		
Non-food, Non-food goals							
Identify hunger	X						
Identify non-hunger eating	X			X			
Timing of meals/snacks	X						
Body image, perceived Fatness	X	X		X			
Good food/bad food Perception	X	X		X			
Frequency of weighing	X	X		X			
Portion Sizes							
Eating out							
Grocery shopping							
Optimum/Practitioner Goals	X	X		X	X		
Assessment							
Determine Level of Risk					X		
Dietetic-Specific Physical Exam*	X: B-vitamin nutrition	X: hair pluckability, lanugo hair	X: parotid gland, Tooth enamel	X		X: mid-arm circum.	X: tenting, edema

ED Matrix p. 2

X axis: If →	Electrolytes Outside of N limits	% Normal Body Wt	Presence of Nutrient-based Lesion(s)	Presence of Non-nutrient-based lesions	Purging; Frequency	Restricting	Bingeing
History: diet, exercise					X	X	
Purging-laxative, emesis, Exercise	X				X	X	
Energy level	X	X			X	X	X
Medical Record Review:							
Labs, consults, Rx, substance Abuse, Hospitalizations,	X	X					
Therapeutic Use of Self	X	X	X		X	X	X
Establish rapport and							
Non-judgmental inquiry							
Client perception of situation, Goals		X			X	X	X
Human Condition →							

ED Matrix p. 3

Y Axis, Then ↓ / Practitioner Actions/Attitudes Interventions						
Schedule follow-up					X	
Biochemical					X	
Refer to Psych If not in Tx					X	
Refer to MD-PCP						
Negotiated Goals: Client:RD						
Wt: Long term	X				X	
Rate of change					X	
Calorie intake	X					
Protein Intake/form	X				X	
Carbohydrate intake/form	X			X		
Fat intake/form	X					
Fiber intake	X					
Fluid intake	X					
Exercise level	X		X			
Vitamin/Mineral Supplements	Individualized			X		
Non-food, Non-food goals						
Identify hunger	X					
Identify non-hunger eating	X					
Timing of meals/snacks	X					
Body image, perceived Fatness						
Good food/bad food Perception				X		
Frequency of weighing	X					
Portion Sizes	X					
Eating out	X					
Grocery shopping	X					
Optimum/Practitioner Goals	Integration, flexibility compromise				acknowledge	acknowledge
Assessment						
Determine Level of Risk						

ED Matrix p. 4

Human Condition →	Readiness for Change	Client Perception of Control	Composition of food intake: Sweets, Fats, Alcohol	Calorie Expenditure: Excess /Inadequate	Predominant Client Concern	Maintenance program
Dietetic-Specific Physical Exam*			B-Vitamins			
History: diet, exercise						
Purging-laxative, emesis, Exercise						
Energy level						
Medical Record Review:						
Labs, consults, Rx, substance Abuse, Hospitalizations,						
Therapeutic Use of Self	X	X				
Establish rapport and						
Non-judgmental Inquiry						
Client perception of situation, Goals						

X axis: If →

Dietetic-Specific Eating Disorder Protocol p.1 of 2

Eating Disorders (Anorexia Nervosa, Bulimia, Compulsive Over-eating)

Rationale for Use of Diet/Nutrition Therapy: To correct problem indicators

1. Restore energy intake appropriate for supporting desired weight status, body composition and exercise level
2. Maintain adequate macro- & micro-nutrient intake
3. Support healthy functioning of the gastro-intestinal system
4. Promote perceptions and skills to support comfortable control over food-related behaviors and weight status

| Human Condition | Practitioners Actions/Attitudes | Outcomes |
| If / Indicators | Then / Interventions | Goals / Corrections |

*Electrolytes; Minerals not Within Normal Limits

If Na, K, Cl high -------->	Minimum 64 oz. Fluid/day	→	Na: 135-145 meq/L
If Na, K, Cl low -------->	>1000 mg Na, K, Cl per day in food		K—3.8-5.5 "
If CO_2 high -------->	Contract for Ø induced vomiting,		Cl—100-108 "
If CO_2, or K low -------->	Contract for Ø laxative use		CO_2—22-34
P, Mg, Ca, Fe			

Very mild exercise ONLY until Normal EKG established—refer to MD.   →   Cessation of purging behaviors

*Body Weight

<85% normal/usual weight ----------> Caloric intake to support weight gain gain of 1-3#/week—(BEE + activity)   ↑   Weight within Normal range at rate of 1-3#/week rate of change

>30% of body weight lost in past 6 mo. -----> Caloric intake to maintain Wt WNL BEE + Activity
N Dx code: 23.006

*Muscle-wasting

Mid-Arm Circumference <80th %ile ----> Protein intake 0.8-1.0 g/kg desired body weight Moderate level regular exercise   →   Body composition improvement as weight normalizes
for age and gender

Females: 20-27% body fat BMI: 20-27
Males: 12-18%   "   BMI: 20-27
Serum Protein   3.5-5.0 g/dl
    "   Total Protein 6.0-8.0 "
Albumin/Globulin Ratio 1.1-2.4

Dietetic-Specific Eating Disorder Protocol, p. 2 of 2

Eating Disorders (Anorexia Nervosa, Bulimia, Compulsive Over-eating)

Human Condition If / Indicators	Practitioners Actions/Attitudes Then / Interventions	Outcomes Goals / Corrections
*Presence of nutrient-based lesions - - - - ->	Dietary adequacy, variety Moderate level, multi-vitamin/mineral - - - - - - - -> Biochemical confirmation	Absence of nutrient-based lesions (re-examine in 2 months)
*Constipation, Bloating, Diarrhea - - - -> Abdominal distention, Flatulence	Fluid intake ≥ 64 oz/day Gradual increase of fiber to 20-30 g/day - - - - - - -> Moderate level exercise program	Regular bowel movements Cessation of symptoms Cessation of purging with laxatives
*Inappropriate food role perception, - - - -> food abuse (bingeing, compulsiveness, control issues,	Client-specific behavioral/perception goals examples: identify hunger/non-hunger eating - - - -> good-bad food perception, frequency of monitoring weight	Verbal expression by client of change, improvement progress toward other goals re: weight, lesions, diet composition, regularity of eating etc.

B. Vignettes from Practice

The domains of The Metaparadigm of Clinical Dietetics can be used to conceptualize circumstances that occur in clinical practice. Below are vignettes from practice. After reading a situation, identify the domain(s) of concern to clinical dietetics, or whether the situation is outside the concerns of clinical dietetics. Situations may represent relationships 1) between two aspects within one domain, such as two factors in the Client Environment which are influencing eating behavior; 2) situations between two domains of the metaparadigm, such a Practitioner's Actions related to a specified Human Condition; 3) situations reflecting a domain of the metaparadigm and a factor outside of clinical dietetics such as the relationship between a Nutraceutical and a pharmaceutical or 4) a concern outside of the practice of clinical dietetics such the menu for the department awards banquet.

Vignettes

Circle your choice of answers (1-6) and fill in the blanks of the answer you select with the name of the Metaparadigm Domain(s) you feel are relevant to the situation.

<u>Metaparadigm Domains</u>

Reference Person

Human Condition

Practitioner Actions/Attitudes

Practitioner Environment

Client Actions/Attitudes

Client Environment

Nutraceuticals

1. **If a dietitian questions a patient's parents about meals the child eats at school and at the grandparent's home, then …**
    1. This represents a within-domain concern of Domain _____.
    2. This represents a between-domain relationship of Domains _____ and _____.
    3. This represents a relationship between domain (s)_____ and a concern outside dietetics.
    4. This represents a concern outside professional boundaries.
    5. This represents a professional concern, but isn't included in a Metaparadigm Domain.
    6. Before deciding I need information about _____.

2. **If a Dietitian is concerned with the ethics of treating obesity when the treatment has such a high failure rate, then …**
    1. This represents a within-domain concern of Domain _____.
    2. This represents a between-domain relationship of Domains _____ and _____.
    3. This represents a relationship between domain (s)_____ and a concern outside dietetics.
    4. This represents a concern outside professional boundaries.
    5. This represents a professional concern, but isn't included in a Metaparadigm Domain.
    6. Before deciding I need information about _____.

3. **If a dietitian collaborates with a primary care provider (MD, PS, NP) to assess the process and outcome of treatment involving diet and drugs, for example use of Redux for appetite control while lowering calorie intake, cycling monthly between on/off Redux and 1200/1500 calories, then …**
    1. This represents a within-domain concern of Domain _____.
    2. This represents a between-domain relationship of Domains _____ and _____.
    3. This represents a relationship between domain (s)_____ and a concern outside dietetics.
    4. This represents a concern outside professional boundaries.
    5. This represents a professional concern, but isn't included in a Metaparadigm Domain.
    6. Before deciding I need information about _____.

4. **If a dietitian performs a physical examination to determine if a patient with bulimia is malnourished even though weight is WNL, then ...**

    1. This represents a within-domain concern of Domain _____.
    2. This represents a between-domain relationship of Domains _____ and _____.
    3. This represents a relationship between domain (s)_____ and a concern outside dietetics.
    4. This represents a concern outside professional boundaries.
    5. This represents a professional concern, but isn't included in a Metaparadigm Domain.
    6. Before deciding I need information about _____.

5. **If a dietitian enrolls in a graduate school program, then ...**

    1. This represents a within-domain concern of Domain _____.
    2. This represents a between-domain relationship of Domains _____ and _____.
    3. This represents a relationship between domain (s)_____ and a concern outside dietetics.
    4. This represents a concern outside professional boundaries.
    5. This represents a professional concern, but isn't included in a Metaparadigm Domain.
    6. Before deciding I need information about _____.

6. **If a dietitian suspects that a client's appetite is being affected by depression, discusses it with the patient and recommends a self-referral to the Psychology Department, then ...**

    1. This represents a within-domain concern of Domain _____.
    2. This represents a between-domain relationship of Domains _____ and _____.
    3. This represents a relationship between domain (s)_____ and a concern outside dietetics.
    4. This represents a concern outside professional boundaries.
    5. This represents a professional concern, but isn't included in a Metaparadigm Domain.
    6. Before deciding I need information about _____.

7. **If a dietitian works in a research facility in the process of determining the need for Boron intake in post-menopausal women, then …**

    1. This represents a within-domain concern of Domain _____.
    2. This represents a between-domain relationship of Domains _____ and _____.
    3. This represents a relationship between domain (s)_____ and a concern outside dietetics.
    4. This represents a concern outside professional boundaries.
    5. This represents a professional concern, but isn't included in a Metaparadigm Domain.
    6. Before deciding I need information about _____.

8. **If a dietitian has an appointment with the hospital administrator, then …**

    1. This represents a within-domain concern of Domain _____.
    2. This represents a between-domain relationship of Domains _____ and _____.
    3. This represents a relationship between domain (s)_____ and a concern outside dietetics.
    4. This represents a concern outside professional boundaries.
    5. This represents a professional concern, but isn't included in a Metaparadigm Domain.
    6. Before deciding I need information about _____.

9. **If a dietitian recommends use of a multi-vitamin/mineral supplement to a patient who reports he drinks a six-pack of beer daily, then …**

    1. This represents a within-domain concern of Domain _____.
    2. This represents a between-domain relationship of Domains _____ and _____.
    3. This represents a relationship between domain (s)_____ and a concern outside dietetics.
    4. This represents a concern outside professional boundaries.
    5. This represents a professional concern, but isn't included in a Metaparadigm Domain.
    6. Before deciding I need information about _____.

10. If a dietitian serves healthy snacks to the hiking club when it is her turn to bring refreshments, then …

1. This represents a within-domain concern of Domain _____.

2. This represents a between-domain relationship of Domains _____ and _____.

3. This represents a relationship between domain (s)_____ and a concern outside dietetics.

4. This represents a concern outside professional boundaries.

5. This represents a professional concern, but isn't included in a Metaparadigm Domain.

6. Before deciding I need information about _____.

Identify the Domains of The Metaparadigm of Clinical Dietetics addressed in articles published in the Journal of the American Dietetic Association. (August, 2001):

1. **If** an article is published titled *Past, present and future of the Food Guide Pyramid* (Davis, Britten and Myers, 2001) **then** this falls within the domain of: _____.

2. **If** an article is titled *Post-diagnosis family adaptation influences glycemic control in women with type diabetes* (Gerstle, Varenne, Contento, 2001), **then** this concerns a relationship between the domains of _____ and _____.

3. **If** an article is titled *Higher carbohydrate intake is associated with decreased incidence of newborn macrosomia in women with gestational diabetes*, (Romon, et al., 2001) **then** this addresses a relationship between the domains of _____ and _____.

(There are no absolute answers. Perspectives on the answers may vary with individuals. The perspective of the author appears in Appendix A.)

C. Linking The Metaparadigm of Clinical Dietetics to a Nutritional Care Process Model

Splett and Myers (Splett & Myers, 2001) describe a model of the nutritional care process as having three components/domains (spheres of activity and influence): a trigger event, the nutrition care process and nutrition-related outcomes. The topics discussed by Splett and Myers describing and characterizing the model and the nutrition care process can be linked to the seven domains of The Metaparadigm of Clinical Dietetics with parallel knowledge topics used to characterize the domains of Metaparadigm of Clinical Dietetics.

The table below illustrates that the proposed Metaparadigm of Clinical Dietetics embraces the Splett and Myers model while noting the areas were perceived as unique to the profession of clinical dietetics.

Linking the Nutrition Care Model to this initially validated global statement of concerns defining Clinical Dietetics serves as theoretical grounding of nutritional care in the practice of clinical dietetics. Nutritional care is defined as a clinical dietetic role. Using the theoretical support of the Metaparadigm of Clinical Dietetics has the potential to support the domains of practice of clinical dietetics in legislation and procedural codes.

Linking a Nutrition Care Model* to the Metaparadigm of Clinical Dietetics

Topics as Links

Nutrition Care Domains (N=3)	Nutrition Care Model Topics	Metaparadigm of Clinical Dietetics Topics	Metaparadigm Domains Domains (N=7)
Trigger Event	Pt. identified as appropriate for nutrition care	Normal and departure from normal appearance of tissues likely to develop nutrient-based lesions	**Human Condition**
	by RD, DT, other health care providers	Interdependent colleagues	**Practitioner Environment**
	General health screening Disease-focused screening	Acceptable laboratory test ranges	**Reference Person**

Linking a Nutrition Care Model* to the Metaparadigm of Clinical Dietetics

Topics as Links

Nutrition Care Domains (N=3)	Nutrition Care Model Topics	Metaparadigm of Clinical Dietetics Topics	Metaparadigm Domains Domains (N=7)
Nutrition Care Process	Indicators of risk, strengths, obstacles, causes, urgency	*Assessment of diet, environment,* health status, mental status, tissue appearance, drug/nutrient interaction, etiologies, nutritional diagnosis	**Practitioner Actions/ Attitudes** **Client Environment**
	Establishes goals and interventions, prescribe, educate, counsel, motivate, refer	*Nutritional goal-setting, nutritional education, counseling, using dietetic protocols, menu writing*	**Practitioner Actions/ Attitudes** **Nutraceuticals**
	Document, establish a record, links	*Documentation/expression of findings using nutritional diagnostic codes and classification system* electronic communications	**Practitioner Actions/ Attitudes** **Practitioner Environment**
Nutrition-Related Outcomes	Direct nutrition outcomes	Acceptable laboratory test ranges	**Reference Person**
	Utilization, cost savings	*Management of clinical dietetic services, measurement of outcomes*	**Practitioner Actions/ Attitudes**
	Self-management	Client lifestyle and food choices, ability and *knowledge for nutritional self-care*	**Client Actions/Attitudes**

(Italics indicate Metaparadigm Topics that practitioners perceived as being unique to clinical dietetics when compared to other health professions or the nutritional sciences.)

*Splett, Patricia and Meyers, Esther, A proposed model for effective nutrition care. J Amer Diet Association. 2001;101:357-363.

It can be seen that components of the nutritional care process link to all seven domains of The Metaparadigm of Clinical Dietetics. In a sense this cross-validates the two. It demonstrates the inclusiveness of The Metaparadigm of Clinical Dietetics: it demonstrates that the nutritional care process includes all domains with which Clinical Dietetics is concerned.

Initial research on The Metaparadigm of Clinical Dietetics also demonstrates agreement with Splett and Meyers regarding the perceived lack of defined roles. Role blurring or delineation influences the practice of clinical dietetics. Practitioners responding to the survey compared their perceptions regarding relevance of knowledge used in clinical dietetics to relevance of the same knowledge to 30 other major professional categories (nurses, physicians, pharmacists, psychologists, social workers, physical therapists, occupational therapists and speech therapists, plus 22 additional variations of professionals with whom they worked.) Clinical Dietitians appear to compare their use of nutritional knowledge with use of similar knowledge by numerous other professions.

The Metaparadigm topics adapted to the Nutrition Care Model were selected from those presented to a randomized national survey sample and the subsequent data base of responses. The number of clinical dietetic respondents to the survey totaled 136. It is important to emphasize that the Metaparadigm Domains represent a proposed template for building our unique body of knowledge as related to medically classified individuals.

Also noted: all topics perceived as unique to clinical dietetics are found in the domains of Practitioner Actions/Attitudes and Human Condition, with use of nutrition diagnostic codes to express the Human Conditions. This points the way to areas of future research in clinical dietetics as clinical dietitians build their body of profession-specific knowledge in various Practitioner Environments.

D. Structuring a Practitioner Environment— Organizing a Clinical Practice

A Clinical Dietitian in private practice or in business may utilize The Metaparadigm of Clinical Dietetics to structure and define her enterprise. For example, a practitioner might ask: Which domains does my business address? Are there any domains purposefully excluded? How do I define the domains to convey the purpose and process of my business?

Brief Example: _____ (name of business) _____

Reference Person:
Normal laboratory biochemistry standards used
Treatment Protocols used

Human Condition:
Referrals for which nutrition problems will be accepted/treated
Nutrient-based lesions examined for, observed

Practitioner Actions/Attitudes:
Accept referrals
Nutritional Diagnosis
Outcomes research
Professional communication

Practitioner Environment:

Business structure, locations, contracts

Standards of Practice

Credentials required

Client Actions/Attitudes:

Readiness

Behavioral changes

Client Environment:

Family support

Insurance

Cultural factors

Nutraceuticals:

Recommendations re: foods, calorie levels, calorie supplements, vitamins, minerals, other nutraceuticals,

Education re: nutrient:drug interactions.

Referrals, resources, but not recommendations re: herbs

IV. Dietetic Education

A. Advanced Level Practice

As a statement of the concerns of the discipline, the Metaparadigm of Clinical Dietetics can be used as a guide for educating practitioners. The seven domains and ninety-four characterizations, or content descriptions of the domains, can serve as an outline during curriculum planning for novices or advanced clinicians. Educators may use the domains of The Metaparadigm of Clinical Dietetics to organize, develop or evaluate the scope of an educational program.(Christy & Kight) (Thompson & Kight)

One means of assessing change in an individual due to education is "before and after" testing. Results can denote changes in knowledge of facts, or change in attitude or perception. Use of the survey instrument for The Metaparadigm of Clinical Dietetics as the means for detecting a change in perception of relevance regarding knowledge that is used by practitioners is one potential avenue for detecting change related to an educational program.

B. Organizing a Curriculum Using the Metaparadigm

A simplified example of a hypothetical curriculum for an advanced degree in clinical dietetics, organized using The Metaparadigm of Clinical Dietetics follows:

Hypothetical Advanced Clinical Dietetic Curriculum Structured by the Metaparadigm of Clinical Dietetics

Reference Person	Human Condition	Practitioner Actions	Practitioner Attitudes	Practitioner Environment	Client Actions/ Attitudes	Client Environment	Nutraceuticals
Knowledge of Appropriate reference values to apply	Nutriologic Person: Physical Examination Procedures for identifying overt nutrient-based lesions	Diagnostic thinking	Collegiality	Institution	Assessing readiness to change	In-patient: Hospital, Rehab, Long term care Out-Patient: HMO, Home care; Community/ Epidemiology	Etiology-driven use of vitamin/mineral supplements
Selection of appropriate biochemical tests	Use of Nutrition Diagnostic Codes	Characteristics of an Expert	Ethical Issues	Clinical experience rotations	Assessing world view of client	Drug/Nutrient interactions	Pharmafoods
	Advanced Nutrio-therapeutics	Therapeutic use of self	Metatheory	Mentoring relationships		National Health Care Policy	Phytochemicals
	Controversies in therapeutics	Counseling/ teaching skills	Professional self-esteem	Funding support for students			Enteral Nutrition-delivery systems
	Metabolic set-points	Creating Tx protocols/ interventions/ critical paths based on Nutritional risk/status		Use of current technology; education methods; electronic medical records			
	PsychoNutriologic Person	Building new knowledge		Admission requirements			
	Cognitive effects of nutritional status	Evaluation of body composition		Other professions' diagnostic codes			
	Interaction of mind & body	Clinical Nutrition examination procedures					

C. From Novice to Expert

To progress from novice to expert (Biesecker, 1999) practitioners may choose to become more diverse or increasingly specialized. Using The Metaparadigm of Clinical Dietetics as a tool for reflection and self evaluation can increase the clarity of such a process. Individuals creating a personal Professional Development Portfolio, or five-year plan, along with the desired continuing education or other activities for reaching the goal(s), may find the following sample and blank tool a stimulus for thought. Of course, these tools may be modified to meet the needs of group work, or of individuals at one point in time or over time.

In the following example Practitioner Actions and Practitioner Attitudes have been divided and portrayed separately to allow for more detail. This concept requires more analysis and clarification. The classification of Expert may possibly be found to coincide with the concept Advanced Level Practitioner.

Novice to Expert Continuum Structured by the Metaparadigm of Clinical Dietetics

	Reference Person	Human Condition	Practitioner Actions	Practitioner Attitudes	Practitioner Environment	Client Actions/ Attitudes	Client Environment	Nutraceuticals
Expert	Investigate Reference Values	Absence of/ Limited Nutritional Collateral Systems	Create new Profession-Specific Knowledge	Altruism Global Thinking	World-Wide Profession	Seek Support for process of change	World/Nation	Herbs/Spices Phyto-chemicals
	Develop technology for investigating reference values	Possible development of Specific disease	Manage clinical Cases Therapeutic Use of Self		National Association; social, Legal expectations	Become educated, skillful self-reliant	Community	Pharmafoods
	Effect of environment on Nutrient needs/use	Inappropriate Food Role Perception/ Abuse	Diagnose Nutritional Problems	Mentor/Empower others—their goals	State/Local Association Social/legal Expectations		Work site	TPN
	Effect of Health/Disease status on nutrient needs/use	Deficiencies, Excesses Balance	Physical examination: identify nutrient-based lesions			RD-Client Joint goal-setting	HMO Outpatient	Vitamins/ Minerals—Oral Preventive/ Restorative
		Deficit in Nutrition Knowledge	Treat Nutritional Problems Diagnosed by others	Internalized ethics Support shared goals Share ideas, feelings	Community Involvement	RD Instructs client what to do	Home Care	
	Effect of non-Disease states on Nutritional status— ex-pregnancy	Impaired Exercise Performance	Counsel	Responsible for having authority; Share knowledge, resources to meet others needs	Colleagues within/outside of dietetics	RD Helps client	Physical Rehabilitation	Calorie Supplementation— Oral
		Inappropriate Dietary Habits	Educate				Long term care Facility	Manufactured Foods
Novice	Age/Gender RDA	Over/Under-weight status	Plan medical Nutritionally Therapeutic menus/guides	Remain in control Tell others; motivate through fear; Take orders, Compartmental thinking	Personal Practice Environment	RD does it for client	Hospital	Foods

Novice to Expert Continuum Structured by the Metaparadigm of Clinical Dietetics

	Reference Person	Human Condition	Practitioner Actions	Practitioner Attitudes	Practitioner Environment	Client Actions/ Attitudes	Client Environment	Nutraceuticals
Expert								
Novice								

Group Evaluation Using Novice to Expert continuum

Self-Evaluation of Practice Level Using the Metaparadigm of Clinical Dietetics and the Novice to Expert Continuum

2/97 Tucson, Arizona
N=8
Total possible Score   = 64.0 (8 scores per individual, maximum of 8 as "expert")
Mean Total Scores   = 41.6 (372 ÷ 8)
Mean Group Practice Level =   5.57 (8 subjects, 8 mean scores across domains)
Range of Domain Scores: 5.0-6.1

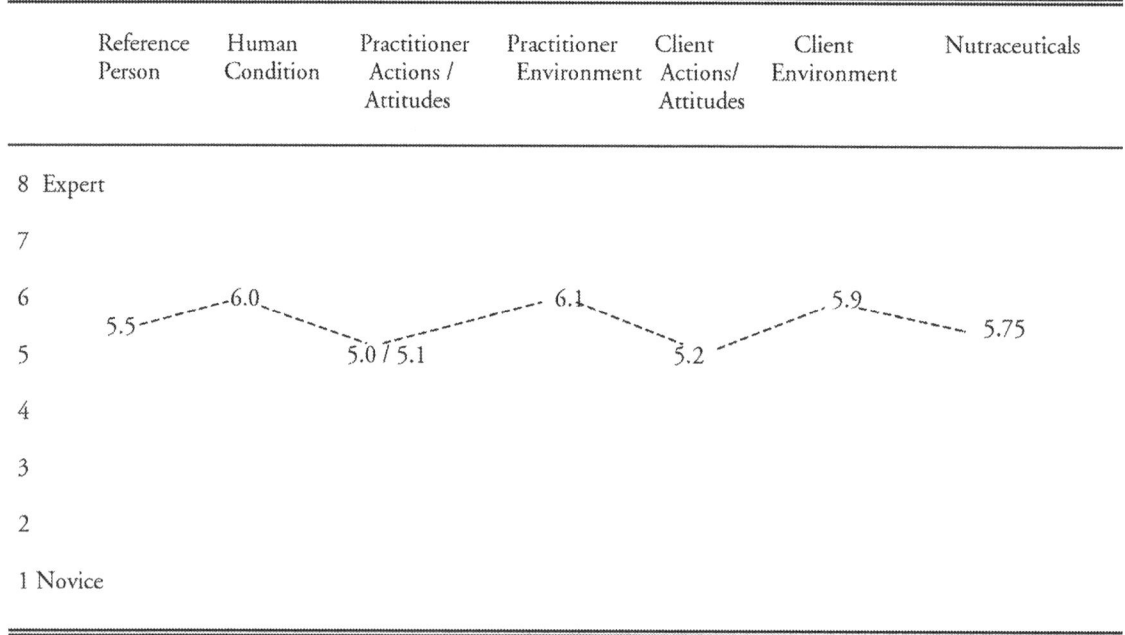

D. Practitioners' Perceptions of Learning Needs

Kein, Johnson and Gates (Kein, et al, J Amer Diet Assoc, 2001) report knowledge and learning needs of 372 RDs reported as part of the pilot test of the Professional Development Portfolio process.

Using the domains of The Metaparadigm of Clinical Dietetics, learning needs may be conceptualized as follows:

Domain of Metaparadigm Of Clinical Dietetics	Category of skill or knowledge needed + Number of RD's listing this need
Reference Person:	Screening parameters, methodology, surveillance (50)
	Nutrition biochemistry (44)
	Other science of food and nutrition (34)
Human Condition:	Disease/Disorder (155)
	Disease Prevention (91)
	Other MNT (88)
	Enteral and Parenteral support (73)
	Diabetes (57)
	Lifecycle (46)
Practitioner Actions:	Research and Grants (135)
	Communication skills: Verbal (130) and written (117)
	Assessment (Nutritional assumed) methodology (80)
	Leadership (47)
Practitioner Attitudes:	Counseling, therapy and facilitation skills (93)
	Critical and strategic thinking (39)
	Team Building, Mentoring, Coaching (37)
Practitioner Environment:	Computer technology (278)
	Marketing (63)
	Career Planning (55)
	Quality Management (53)
Client Actions	
Client Attitudes	
Nutraceuticals:	Supplemental nutrients, complementary, and herbal therapies (130)
	Pharmacological/drug/nutrient/herbal interaction (63)

The highest perceived needs of RD's are from the professional domains of

1) Practitioner Environment (Computer technology—278)
2) Practitioner Actions (Communication skills—247) (Research and Grants—135)
3) Human Condition (Disease—155)
4) Nutraceuticals (Supplements—130).

If another study is done at a later date or with a different group, domain comparisons can be one way to identify changes in global perception of educational needs.

* The item from this study identified as "Food Safety, HACCP and sanitation" is felt to fall outside the domains of concern to Clinical Dietetics. This item would fall within the domain of Practitioner Environment in the proposed expansion of this model to The Metaparadigm of Dietetics.

V. Dietetic-Specific Research

A. Clinical

Building Dietetic-Specific Clinical Knowledge

Clinical practice is the potential research site for clinical dietetics. (Dwyer, 1997) Practice is the laboratory of the practitioner. Research may investigate a.) within-domain questions, b.) between-domain questions, or c.) relationships of a clinical dietetic domain to other health professions or nutritional science.

Examples of clinical research questions conceptualized using domains from The Metaparadigm of Clinical Dietetics include:

A. Within-Domain question:

> What skills must a client gain to progress from goal-setting with the clinical dietitian to become self-reliant in goal-setting? (one Client Action/Attitude to another)

B. Between-Domain questions:

> How does Practitioner Attitude relate to Client Attitude?
>
> > How does the practitioner's *ability to empathize* relate to client *willingness* to keep follow-up appointments.
>
> How does use of a Nutraceutical affect the Human Condition?
>
> > How does the *level of folic acid intake* from reported *dietary* sources relate to *degree of hyperhomocysteinemia* and/or *risk* of heart disease?

C. Clinical Dietetic domain and domains of other professionals and scientists:

> How does the Practitioner Action of *nutritional assessment* relate to use of an assessment tool created by a Nutritional Scientist? Consider that an assessment tool may be used in assessing 1) physical attributes such as body composition, 2) mental attributes such as knowledge level, or 3) emotional attributes such as readiness for change.

> How does the *readiness score* relate to *degree of success* for clients in weight loss program? (Success defined as behavior change or pounds lost or length of time weight loss maintained)

B. Research Process

Using the Metaparadigm of Clinical Dietetics in defining the clinical research process and demonstrating the four levels of knowledge

Example: Defining the Research:

Metaparadigm Global Concepts:

>   Practitioner Action

>   Client Actions

>   Nutraceuticals

Paradigm Concepts:

>   Counseling by Clinical RD

>   Client's behavior

>   Food choices

>   Exercise level

Model/Assumptions:

>   The Clinical Dietitian's role is to influence client food selection behavior.

>   Counseling by the Clinical Dietitian influences clients to change behavior.

>   Client's changes in food choices result in weight change.

Paradigm Relational Statement:

Following counseling by the clinical dietitian the *client will* select foods differently than before counseling, *while keeping* their exercise level constant. *This will result in* weight loss.

Theory/Hypothesis:

Following three appointments (one one-hour and two half-hour appointments, scheduled over a four weeks period), during which food selections were discussed, and including a request to not alter exercise or activity level, the clients of the dietitians at __(site) __will demonstrate a change of food selection by:

1) a verbal and recorded description of behavior change including food types selected and amounts consumed

2) a verbal and recorded report of no change in activity level

3) This will result in a measured weight loss of at least 1 pound of body weight each week.

Science

Initial N= 4
1 disqualified: did not attend counseling appointments
Final N= 3 subjects

Results and findings:

At the finish of the three counseling sessions

> Clients reported the following behavior changes and measurement indicated:

Subject 1: Ate less candy

> Lost 1.5 lbs.

> Kept walking same routine 1 mile/day

Subject 2:   Ate chicken instead of hamburgers

Ate only 2 pieces instead of 3

Baked chicken instead of frying it

Behavior reported at frequency of three times in the week

Lost 0.5 lbs.

Walked from car to parking lot—no change

Subject 3:   Ate 3 meals per day, instead of 2

Ate no snacks

Lost 2 lbs.

Went to the gym once per week, as usual

Statistical description of data

Mean Wt. loss = 4lbs/3 = 0.75 lbs.

Interpretation of data

1 disqualified; did not attend appointments

3 changed eating habits

1 lost < 1 #

Conclusions drawn from data

Counseling by an RD and change of food selection behavior with no change in activity level resulted in weight loss of 0.5 to 2.0 lbs. in a 4-week period.

Evaluation of Science utilized:

Did the theory describe or predict what actually happened?

Were the measures valid?

Were assumptions complete, accurate?

Were any changes needed if study were repeated?

C. Education

Building Knowledge Regarding Clinical Dietetic Education—A Conceptual Model

Education of practitioners is another area for discipline-specific research, which can be based on the following assumptions:

> The Metaparadigm of Clinical Dietetics is an organizing structure for the body of knowledge of clinical dietetics.

> The purpose of undergraduate and graduate educational programs is to educate practitioners for practicing clinical dietetics utilizing the body of unique and shared knowledge identified as necessary to practice the profession.

> Such educational programs can be organized within the framework of the Metaparadigm of Clinical Dietetics.

> The Metaparadigm of Clinical Dietetics can be an organizing structure for research in clinical dietetic education.

Following is a conceptual model of educational programs for clinical dietitians grounded in the work of Christy and Kight (1993), Thompson and Kight (1990), and Leyse and Kight (1993) utilizing the metatheory of The Metaparadigm of Clinical Dietetics to structure/organize the educational research process.

From the interactions of cell contents in the model will emerge the measures of the educational program. These measures will constitute the science, or data, producing knowledge concerning the education of clinical dietitians. Using this model will support clear communication regarding the research.

A Conceptual Model for Building Knowledge Regarding Clinical Dietetic Education

The three dimensions of the model include the 1) characterization of Practitioner Environment, 2) the characterization of the Student Practitioner Actions/Attitudes and the 3) The characterization of the coursework organized using the domains of the Metaparadigm of Clinical Dietetics All attributes on each axis must be quantifiable (measurable). For example, an environmental characteristic, finances, may be defined as costs to be paid by the department, tuition to be paid by the student, stipends or benefits provided to the students, faculty salaries, etc. Measures may be descriptive statistics, such as "Present/Absent", a frequency of occurrence, number of contacts, attitude assessment, a degree or % of occurrence, a degree of change, test scores, time to completion, a mean, correlations resulting from a factor analysis, correlations and "R" in a causal model, etc.

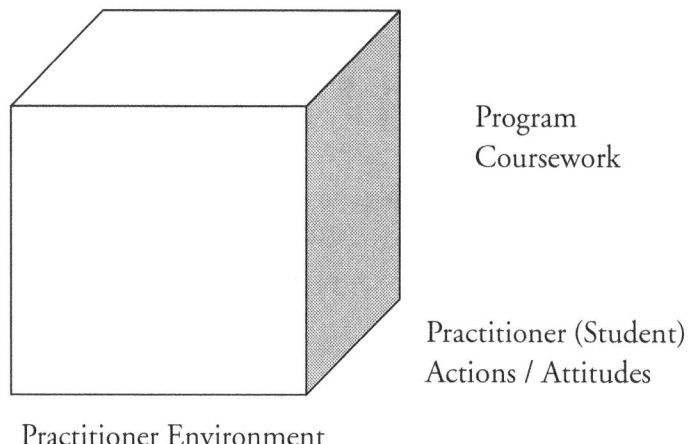

Program
Coursework

Practitioner (Student)
Actions / Attitudes

Practitioner Environment

Practitioner Environment

Practitioner environment refers to characterization of the educational program. Numerous other characteristic attributes could be assigned to the cells of the cube model of the theory. Especially important are attributes that are felt to be influential in the outcome of the educational process; attributes being investigated as part of the research.

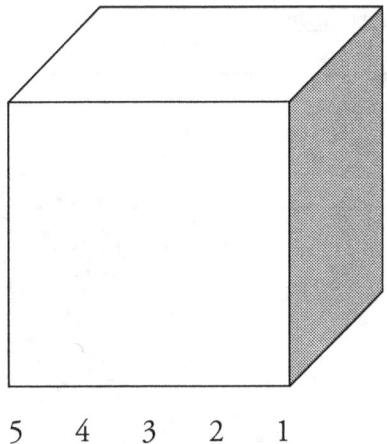

5   4   3   2   1

5. Education Improvement Research
4. Faculty
3. Finances
2. Locations
1. Credentials/Accreditation

5. Education Improvement research
      Method outcomes
      Therapeutic use of self/caring

4. Faculty
      Qualifications
      Roles—mentor, educational research, clinical research
      Relationships—student, colleagues, community

3. Finances
      Cost of program
      Student support

2. Program site
      Residencies—organizing site, affiliation sites
      Distance learning, telecommunication

1. Credentials/Accreditation
      Western Region
      American Dietetic Association

Practitioner (Student) Actions/Attitudes

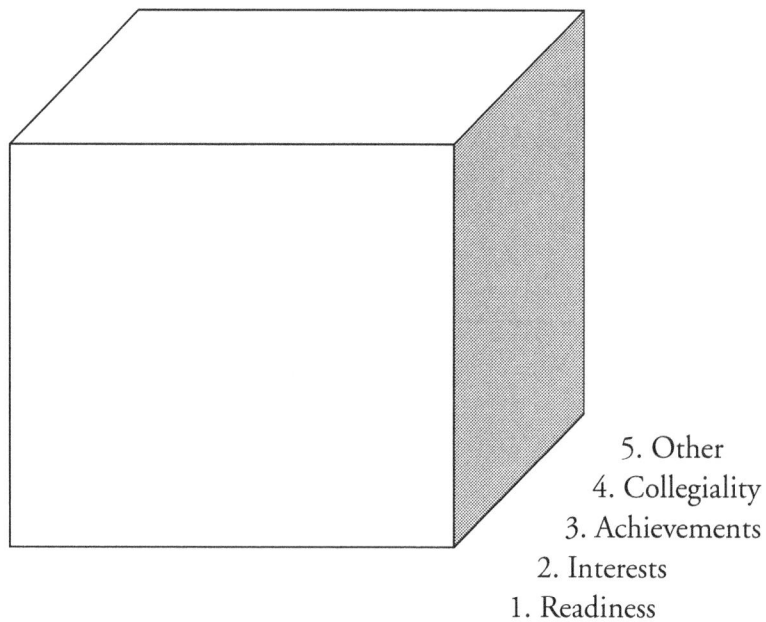

5. Other
4. Collegiality
3. Achievements
2. Interests
1. Readiness

1.   Readiness—experience, mind-set, world view

2.   Interests—plans goals

3.   Achievements—cases, lesion work-ups, treatment matrices, publications, theories

4.   Collegiality—relationships with peers, faculty, community contacts, clients

5.   Other—program-specific

Program Course work-Structured by The Metaparadigm of Clinical Dietetics

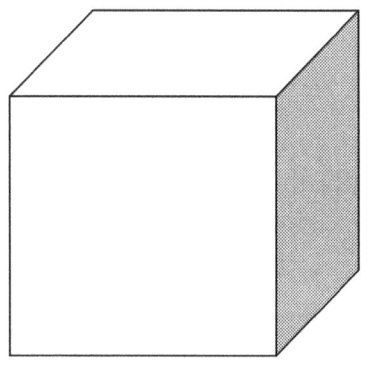

7.   Nutraceuticals

6.   Client Environment

5.   Client Actions/Attitudes

4.   Practitioner Environment

3.   Practitioner Actions/Attitudes

2.   Reference Person

1.   Human Condition

1. Human Condition
   Nutriologic Person
       Dietetic-Specific Diagnostic Codes
       Dietetic-Specific Physical Examination
       Overt nutrient-based lesions
   Psycho-nutriologic Person
       Interaction of Mind/Body (nutritional status)
       Nutriodynamics: Nutrients → Mental functioning
       Nutriokinetics: Cognitive functioning → Nutritional Status

2. Reference Person
   Selection of appropriate biochemical tests
   Knowledge of appropriate reference values to apply

3. A. Practitioner Actions
   Diagnostic thinking
   Therapeutic Use of Self
   Create critical paths, treatment protocols based on assessment of
   Nutritional risk, best practice
   Knowledge development: clinical and educational research

3. B. Practitioner Attitudes
   Ethical issues, values clarification
   Profession-specific theory; meta-theory, middle range theory, current dietetic
       theory, borrowed theory
   World View

4. Practitioner Environment
   Current technology—teaching tools (distance education, computer programs,
       diagnostic codes of other professions), electronic medical records

5. Client Actions/Attitudes
   Assessing client readiness to change
   Assessing client world view

6. Client Environment
   Epidemiological methods, research
   Health care system, policy

7. Nutraceuticals
    Etiology-driven use of vitamin/mineral supplements
    Pharmafoods

Unit interactions using the above characterizations: 5 X 5 X 7 yields a minimum of 175 interactions for potential definition, characterization, development, measurement and assessment

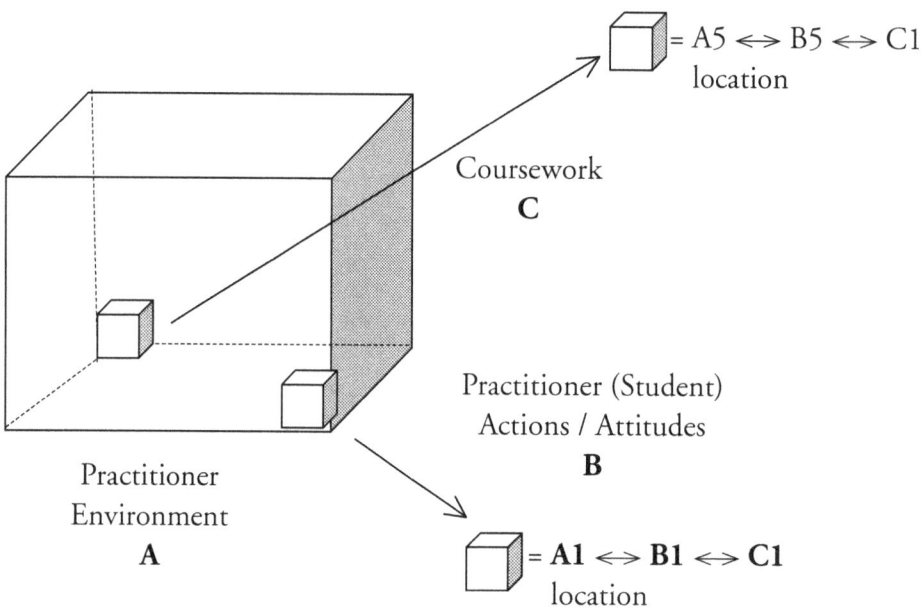

A. Practitioner Environment
    1. Accreditation
    2. Faculty
    3. Finances
    4. Location
    5. Educational Improvement Research

B. Practitioner Actions/Attitudes
    1. Readiness
    2. Interests
    3. Achievements
    4. Collegiality
    5. Other

C. Program Coursework
    1. Human Condition
    2. Reference Person
    3. Practitioner Actions/Attitudes
    4. Practitioner Environment
    5. Client Actions/Attitudes
    6. Client Environment
    7. Nutraceuticals

Following are three examples of the minimum of 175 interactions for potential definition, characterization, development and assessment:

Example I.

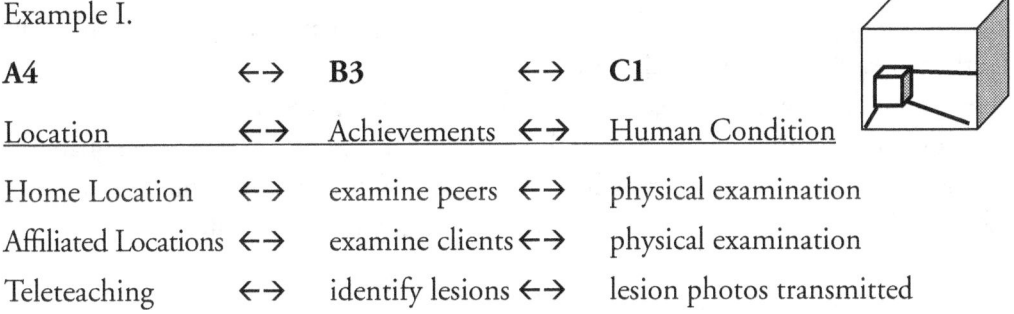

A4	←→	**B3**	←→	**C1**
Location	←→	Achievements ←→		Human Condition
Home Location	←→	examine peers	←→	physical examination
Affiliated Locations	←→	examine clients	←→	physical examination
Teleteaching	←→	identify lesions	←→	lesion photos transmitted

Research: Which locations promote speed and quality of achievement in the area of physical examination of nutrient-based lesions.

Example II.

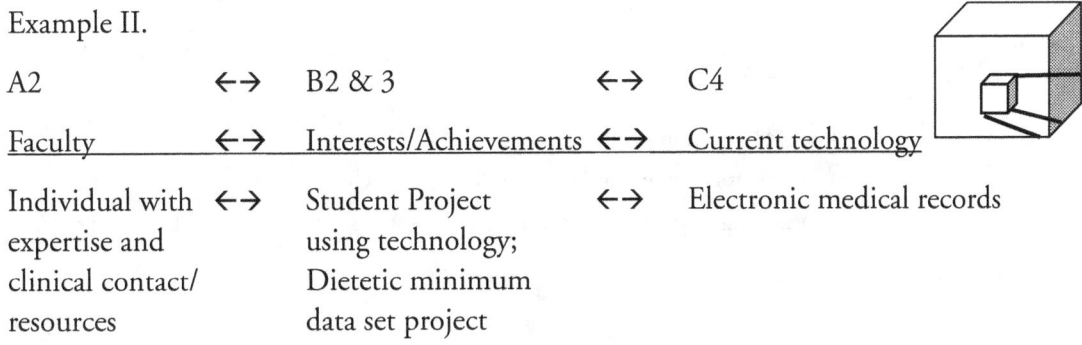

A2	←→	B2 & 3	←→	C4
Faculty	←→	Interests/Achievements ←→		Current technology
Individual with expertise and clinical contact/ resources	←→	Student Project using technology; Dietetic minimum data set project	←→	Electronic medical records

Research: Are electronic medical records a useful technology for educating students?
        Are students able to create minimum data sets from available electronic medical records?

Example III.

A5   ←→ B1   ←→ B3

Education   ←→ Student Readiness ←→ Individual Practitioner(Student Achievement)

Improvement   Measures of   Score on Registration examination

Knowledge   Readiness   Completion of program within expected time frame

Research: Which assessment indicators predict and promote student readiness to enter and successfully complete program

D 1) Describing Profession-specific literature

Clinical Dietetic research described by professional domains of concern could easily be compiled and compared with other research in those domains. Reviews of domains reported in published research can indicate where research is needed. Research devoted to the domains of The Metaparadigm of Clinical Dietetics could be compared with research by Clinical Dietetic Practitioners in collaboration with nutritional scientists or with other health professionals. Selected articles published in the June, 2001 issue of *The Journal of the American Dietetic* Association can be classified as follows:

Metaparadigm Domain	Titles of Research Articles in *J Amer Diet Assoc*, June, 2001.
Human Condition:	*The effects of antiretroviral protease inhibitors on serum lipid levels in HIV-infected patients *Nutrient intake of infants hospitalized with lower respiratory tract infections
Reference Person:	*The fifth edition of the Dietary Guidelines for Americans: Lessons learned along the way
Practitioner Actions:	*Considerations in planning vegan diets: Children *Considerations in planning vegan diets: Infants *Developing actionable dietary guidance messages: Dietary fat as a case study

Metaparadigm Domain	Titles of Research Articles in *J Amer Diet Assoc*, June, 2001.
Practitioner Environment:	*Reliability and validity of the Child and Adolescent Trial for Cardiovascular Health (CATCH) Food Checklist: A self-report instrument to measure fat and sodium intake by middle school students *Screening method evaluated by nutritional status measurements can be used to detect malnourishment in chronic obstructive pulmonary disease *Learning needs and continuing professional education activities of Professional Development Portfolio participants. (see also p. 40) *Tips for contract negotiations and establishing MNT rates
Client Environment:	*Focus groups indicate vegetable and fruit consumption by food-stamp-eligible Hispanics is affected by children and unfamiliarity with non-traditional foods *A super-sized problem: Restaurants piling on the food
Nutraceuticals:	*Ginseng does not enhance psychological well-being in healthy, young adults: Results of a double-blind, placebo-controlled, randomized clinical trial

D 2) Description of Key Trends

Selected key trends identified in the 2002 Environmental Scan (1) which are predicted to affect the dietetics profession, as organized within the structure of the Metaparadigm of Clinical Dietetics (2).

The Metaparadigm of Clinical Dietetics refers to the seven global, abstract concepts that describe the phenomena of concern to the profession of clinical dietetics and are the guiding concepts which encompass the less abstract levels of professional knowledge. (See Appendix for original survey re themes for which RD's reported their perceptions The definition of each domain, or concept of concern, is one way to define that discipline or profession and differentiate it from other groups. These other groups may be other health professionals or nutritional scientists. (2).

The 2002 Environmental Scan falls within the domains of Practitioner Environment and Client Environment. However, themes within these environments reflect domains of the body of knowledge of dietetics defined by the seven domains of the Metaparadigm of Dietetics

Reference Person	Human Condition	Practitioner Actions/ Attitudes	Practitioner Environment	Client Actions/ Attitudes	Client Environment	Nutraceuticals

Reference Person: (*Reference Person refers to the theoretical, statistically derived individual representative of the reference population, for example the reference "infant 0.5-1.0 years old" referred to in the Recommended Dietary Allowances. It includes the assumption of a defined criteria of selection and assumes the user is informed regarding the essential details of the derivation. When reference values are used in evaluation or interpretation, it is acknowledged that health and disease are relative, not absolute states.*)

Key Trends:

New information and understanding about the role of diet in health care and chronic health problems, allergies, medical/nutrient interactions

Dietary Reference Intakes—use, expansion

Human Condition: (*Human Condition: For clinical dietitians Human Condition refers to the nutritional status of individuals in a state of health or with nutritional problems. The scientifically derived reference status is compared with observed departures from "Normal" status to assess the human condition of clients. Such assessment gives direction to the practitioner providing nutritional care to clients.*)

Key Trends

Nutriologic Person

Increasing cultural and ethnic diversity presenting to RD's

Definition of obesity as public and personal health problem

Increased incidence of diabetes, pre-diabetes

Genetic testing personalizing diagnosis of nutritional status and needs

Aging of the US population and increase of chronic diseases, need for hospice and palliative care

PsychoNutriologic Person

Cognitronics—nutritional aspects of brain function, alertness and mood

Practitioner Actions/Attitudes: (*Actions refers to behaviors engaged in or purposefully refrained from, relative to the practice of the profession or to professional development. Attitude refers to intra-personally based thoughts and feelings about aspects of the clinical dietitian's professional role enactment that are elicited by situational cues. They may be explicit and willfully affect professional behavior, or implicit (unstated, unknown, subconscious or unconscious) and involuntarily affect behavior, not being under the influence of the will. In this instrument attitude includes, but is not limited to ability, aptitude, beliefs, decisions, emotions, ethics, ideas, knowledge, morals, opinions, preferences, thoughts, values, will and world view.*)

Key Trends

Increased personalization of nutrition and dietary recommendations

Need to overcome impressions that the field takes a narrow view of health

Being informed regarding vitamins, minerals, herbal and botanical supplements

Increased accessibility and use of email, online and telephone services

Advocacy for the public as regulations and standards are developed for new technology, food processors, food manufacturers, food service establishments, federal food safety regulations

Changes in purpose, function of organizations

Practitioner Environment: *(Practitioner Environment refers to the complex social and physical circumstances in which clinical dietetics is practiced. It includes relationships with other professionals, the local and national organization, the prevailing political and social milieu, the scientific knowledge, the state of technology available and profession-specific tools.)*

Key Trends

Growth in relevant science and technology—genetics, biotechnology, pharmacogenomics, imbedded smart-chips,

Alternative sources of nutrition information and education available to the public: the world wide web, food producers, PhD nutritionists, alternative medicine practitioners, Physician Nutrition Specialists, personal trainers, private chefs

Increased need for provision of gerontological and pediatric nutritional care

Potential for additional diseases and disorders covered by Medicare for Medical Nutrition Therapy, addressing of nutritional care issues by private insurers

Expanding e-commerce

Potential for expanded scope of practice

New work venues: (neonatal ICU, retirement communities, home health, corporate programs, private client bases, informatics

Client Actions/Attitudes: *(Client refers to any individual or groups that present to a clinical dietitian for nutritional services. Attitudes: refers to intrapersonal characteristics and processes of a client. Attitudes may be explicit and willfully affect client behavior or implicit (unstated, unknown, subconscious or unconscious) and involuntarily affect behavior, not being under the influence of the will. In this instrument attitude includes, but is not limited to ability, aptitude, beliefs, decisions, desires, emotions, ethics, ideas, knowledge, morals, opinions, preferences, thoughts, values, will and world view.)*

Key Trends

Greater public interest and education in nutrition, food safety and functional foods

Decreases in time for food preparation; more meals obtained outside the home

Increased interest and need regarding privacy of confidential health care

Client Environment: *(Client Environment refers to the complex social and physical circumstances surrounding a client who receives nutritional interventions from a clinical dietitian. It includes influences of health status and medications, family and associates, food available, work, finances, cultural influences, the marketplace and self care skills.*

Key Trends

Need for food security

Rising health care costs, need for access to nutritional care by RD's

Need for knowledge re nutritional risk; global standards

Workplace and schools as potential sites for obtaining nutrition information and healthful foods

Increased use of electronic communication with health care providers

Nutraceuticals: *(Nutraceuticals refers to any substance that can be considered to be a food or a component of a food that affects health, including the prevention and treatment of disease. Such products range from all natural, processed, created/engineered, manufactured foods, designer food, functional foods, phytochemicals, isolated nutrients supplements, chemopreventive agents and pharmafoods (Nutraceutical initiative, 1992), (Position Paper of ADA, 1995).*

Key Trends

The distinction between food and medicine is blurring; increased importance of functional foods and designed foods

Genetically altered foods—immune system stimulation, drug delivery, nutritional value enhancement

References

1) Key Trends affecting the dietetics profession and the American Dietetic Association. J Amer Diet Assn. 102: 12. S1820-S1839.

2) Leyse, RL. Perceptions of the Metaparadigm of Clinical Dietetics: A Conceptual Delineation of Phenomena Relevant to the Discipline. UMI Dissertation Services. Ann Arbor Michigan. 1998.

VI. A Metaparadigm of Dietetics?

The following table illustrates the fit of topics traditionally considered non-clinical (for example, administrative dietetics) into the domains initially utilized to propose the Metaparadigm of Clinical Dietetics. It demonstrates that the seven domains are inclusive of topics as diverse as dietetic education, public health nutrition, school food service, dietitians in business, food service management, dietetic research and sensory evaluation.

A more formal and inclusive study of this area could validate the unification of Dietetics under one Metaparadigm for the structure of the entire body of knowledge.

A Metaparadigm of Dietetics?—the fit of diverse practice areas within the domains of a profession-wide Metaparadigm

Journal of the American Dietetic Association Domains Or DPGs	J. American Dietetic Association Topics (abbreviated titles from articles in JADA, 2001	Metaparadigm of Clinical Dietetics Domain Characterizations	Metaparadigm of Clinical Dietetics Domains	Fit within Metaparadigm of Dietetics Domains
Food and Culinary Professionals	Use of Functional Foods; Menu Analysis	Appropriate Use of Nutraceuticals Functions of Nutraceuticals Effects of Nutraceuticals	Nutraceuticals	Nutraceuticals
School Nutrition Service	Important beliefs of parents	Care-giver who is responsible and knowledgeable re: nutritional needs	Client Environment	Client Environment
Dietitians in Business and Communications	What's cooking at ISU … TV show	Links with business & industry Nutrition Education Networks	Practitioner Actions/ Attitudes	Practitioner Actions/ Attitudes
Management in Food and Nutrition Systems	Convenience vs scratch … Customer Satisfaction	Management of Clinical Dietetic Services	Practitioner Actions/ Attitudes	Practitioner Actions/ Attitudes
Nutrition Education for the Public	Impact of computer-assisted instruction	Computer programs re: nutrition and/or diet Electronic communication capability	Practitioner Environment	Practitioner Environment
Management in Food and Nutrition Systems	The Hazard Analysis Critical Control Points … Model … A tool in forensic dietetics	Tools for assessment of … behaviors of individuals Participation in clinical dietetic research as principal investigator Food-borne environmental Toxins	Practitioner Environment Practitioner Actions/ Attitudes Client Environment	Practitioner Environment Practitioner Actions/ Attitudes Client Environment
Nutrition educators of practitioners	Comparing difference of diagrams … multimedia … videos for dietetic education	Computer programs re nutrition and/or diet Nutrition education materials	Practitioner Environment	Practitioner Environment
Nutrition Entrepreneurs	Attitudes & perceptions … working women … non-diet approach … weight management	Ability, Knowledge for nutritional self-care	Client Actions/Attitudes	Client Actions/Attitudes

A Metaparadigm of Dietetics?:—the fit of diverse practice areas within the domains of a profession-wide Metaparadigm

Journal of the American Dietetic Association Domains Or DPGs	J. American Dietetic Association Topics (abbreviated titles from articles in JADA, 2001	Metaparadigm of Clinical Dietetics Domain Characterizations	Metaparadigm of Clinical Dietetics Domains	Fit within Metaparadigm of Dietetics Domains
Public Health Nutrition	Have meal patterns of children changed in ... two decades?	Food choices Caregiver who is responsible ... knowledgeable	Client Actions/Attitudes	Client Actions/Attitudes
Research Dietetic Practice Group	Comparison of bioimpedance methods ... detection of body Cell mass change in HIV infection	Nutritional science laboratory methods	Reference Person	Reference Person
Food and Culinary Professionals	Goat has sensory evaluation ratings similar to beef, pork, lamb	Food preferences Food choices	Client Actions/Attitudes	Client Actions/Attitudes
Nutrition education for the public	Nutrition education preferences of limited resource African American youth	Knowledge for nutritional self-care World view re health and nutrition	Client Actions/Attitudes	Client Actions/Attitudes

VII. Summary of Applications and Recommendations

Summary

The Metaparadigm of Clinical Dietetics is the most abstract level of knowledge used by Clinical Dietitians. The seven concepts of the Metaparadigm are Reference Person, Human Condition, Practitioner Actions/Attitudes, Practitioner Attitudes, Client Actions/Attitudes, Client Environment and Nutraceuticals.

The metatheory of The Metaparadigm of Clinical Dietetics can function in many roles:

It can structure treatment matrices defining best practice.

It provides a means to conceptualize concerns and circumstances in daily practice, one format being vignettes from practice.

It can be used in expressing the nutritional care model and other professional models at the abstract level.

It can form the structure for defining a private practice or other clinical dietetic business.

It can be used to conceptualize, plan and evaluate individual and institutional education of practitioners.

It can structure research in clinical dietetic practice, knowledge building for clinical dietetics, program evaluation and globally conceptualizing clinical dietetic theory.

It can be used to organize and track professional literature and publications.

These functions have potential for organizing profession-specific knowledge by providing a new way to conceptualize the knowledge used and needed.

Recommendations

1 ... that Clinical Dietetic research plans and reports include a conceptual statement describing the domains and relationships addressed, described, elucidated, contributed by the research.

2 ... that education of Clinical Dietetic Practitioners include information regarding the Metaparadigm of Clinical Dietetics and the organizing capabilities it brings to knowledge used by the profession.

3 ... that students and practitioners be conscious of when and how they are using the domains of concern to the profession, when they are in the area of knowledge considered unique and when they are in the area of shared knowledge.

4 ... that the domains of the Metaparadigm of Clinical Dietetics be used as a guideline for practitioners as they create their professional continuing education plans.

5 ... that educational programs use the 7 domains of the Metaparadigm of Clinical Dietetics and the three-dimensional cube model of education for program planning and evaluation.

6 ... that additional validation studies be performed to test this theory with a larger and more inclusive sample of practitioners.

7 ... that, after careful consideration, additional knowledge topics be added to the characterizations of the 7 domains as the profession changes.

8 ... further elucidation of the fit of non-clinical dietetic topics within the seven domains, creating the Metaparadigm of Dietetics.

APPENDICES

Appendix A: Vignettes from Practice

Author's Perspective Regarding Client Vignettes

1. **If a dietitian questions a patient's parents about meals the child eats at school and at the grandparent's home, then ...**
    1. This represents a within-domain concern of Domain _____.
    2. This represents a between-domain relationship of the Domains of <u>Practitioner Actions and Client Environment.</u>
    3. This represents a relationship between domain (s)_____ and a concern outside dietetics.
    4. This represents a concern outside professional boundaries.
    5. This represents a professional concern, but isn't included in a Metaparadigm Domain.
    6. Before deciding I need information about _____.

2. **If a Dietitian is concerned with the ethics of treating obesity when the treatment has such a high failure rate, then ...**
    1. This represents a within-domain concern of Domain _____.
    2. This represents a between-domain relationship of domains <u>Practitioner Attitude and Human Condition</u>
    3. This represents a relationship between domain (s)_____ and a concern outside dietetics.
    4. This represents a concern outside professional boundaries.
    5. This represents a professional concern, but isn't included in a Metaparadigm Domain.
    6. Before deciding I need information about _____.

3. **If a dietitian collaborates with a primary care provider (MD, PS, NP) to assess the process and outcome of treatment involving diet and drugs, for example use of Redux for appetite control while lowering calorie intake, then …**

   1. This represents a within-domain concern of Domain _____.

   2. This represents a between-domain relationship of Domains <u>Practitioner Actions, Client Actions and Human Condition.</u>

   3. This represents a relationship between domain (s) <u>Practitioner Actions, Client Actions.</u> and a concern outside of dietetics. (a Pharmaceutical)

   4   This represents a concern outside professional boundaries.

   5. This represents a professional concern, but isn't included in a Metaparadigm Domain.

   6. Before deciding I need information about _____.

4. **If a dietitian performs a physical examination to determine if a patient with bulimia is malnourished even though weight is WNL, then …**

   1. This represents a within-domain concern of Domain _____.

   2. This represents a between-domain relationship of Domains <u>Practitioner Actions, Reference Person and Human Condition.</u>

   3. This represents a relationship between domain (s)_____ and a concern outside dietetics.

   4. This represents a concern outside professional boundaries.

   5. This represents a professional concern, but isn't included in a Metaparadigm Domain.

   6. Before deciding I need information about _____.

5. **If a dietitian enrolls in a graduate school program, then …**

   1. This represents a within-domain concern of Domain _____.

   2. This represents a between-domain relationship of Domains _____ and _____.

   3. This represents a relationship between domain (s)_____ and a concern outside dietetics.

   4. This represents a concern outside professional boundaries.

   5. This represents a professional concern, but isn't included in a Metaparadigm Domain.

   6. Before deciding I need information about : the subject matter of the graduate program in which the dietitian was enrolling.

6. **If a dietitian suspects that a client's appetite is being affected by depression, discusses it with the patient and recommends a self-referral to the Psychology Department, then ...**

    1. This represents a within-domain concern of Domain _____.

    (2.) This represents a between-domain relationship of Domains <u>Practitioner Actions, ClientAttitude and Human Condition.</u>

    3. This represents a relationship between domain (s)_____ and a concern outside dietetics.

    4. This represents a concern outside professional boundaries.

    5. This represents a professional concern, but isn't included in a Metaparadigm Domain.

    6. Before deciding I need information about _____.

7. **If a dietitian works in a research facility in the process of determining the need for Boron intake in post-menopausal women, then ...**

    1. This represents a within-domain concern of Domain _____.

    (2.) This represents a between-domain relationship of Domains <u>Practitioner Actions and Reference Person.</u>

    3. This represents a relationship between domain (s)_____ and a concern outside dietetics.

    4. This represents a concern outside professional boundaries.

    5. This represents a professional concern, but isn't included in a Metaparadigm Domain.

    6. Before deciding I need information about _____.

8. **If a Dietitian has an appointment with the hospital administrator, then ...**

    1. This represents a within-domain concern of Domain _____.

    2. This represents a between-domain relationship of Domains _____ and _____.

    3. This represents a relationship between domain (s)_____ and a concern outside dietetics.

    4. This represents a concern outside professional boundaries.

    5. This represents a professional concern, but isn't included in a Metaparadigm Domain.

    (6.) Before deciding I need information about: The purpose of the meeting.

9. **If a dietitian recommends use of a multi-vitamin/mineral supplement to a patient who says he drinks a six-pack of beer daily, then ...**

    1. This represents a within-domain concern of Domain _____.

    (2.) This represents a between-domain relationship of Domains <u>Practitioner Actions, Client Actions, and Nutraceuticals.</u>

    3. This represents a relationship between domain (s)_____ and a concern outside dietetics.

    4. This represents a concern outside professional boundaries.

    5. This represents a professional concern, but isn't included in a Metaparadigm Domain.

    6. Before deciding I need information about _____.

10. **If a dietitian serves healthy snacks to the hiking club when it is her turn to bring refreshments, then ...**

    1. This represents a within-domain concern of Domain _____.

    2. This represents a between-domain relationship of Domains _____ and _____.

    3. This represents a relationship between domain (s)_____ and a concern outside dietetics.

    (4.) This represents a concern outside professional boundaries.

    5. This represents a professional concern, but isn't included in a Metaparadigm Domain.

    6. Before deciding I need information about _____.

Identify the Domains of The Metaparadigm of Clinical Dietetics addressed in articles published in the Journal of the American Dietetic Associaton. (August 2001):

1. **If** an article is published titled *Past, present and future of the Food Guide Pyramid* **then** this falls within the domain of: <u>Practitioner Environment</u>

2. **If** an article is titled *Post-diagnosis family adaptation influences glycemic control in women with type diabetes,* **then** this concerns a relationship between the domains of <u>Client Environment and Human Condition</u>

3. **If** an article is titled *Higher carbohydrate intake is associated with decreased incidence of newborn macrosomia in women with gestational diabetes,* **then** this addresses a relationship between the domains of <u>Nutraceuticals and a Human Condition.</u>

Appendix B: Psychometric Methods

	Summary of Psychometric Methods Used in this Study
Sorting and Clustering	Reducing literature to knowledge topics Reducing interviews to knowledge topics Clustering knowledge topics into domains
Content Validity Index	Expert Panel questionnaire
Qualitative Interviews	Discovery from practitioners; practice-based compared to literature-based concerns Reducing transcribed data to themes, categories
Pilot Testing of Instrument	Test with practitioners unfamiliar with concepts Test burden on respondents (time)
Descriptive statistics Frequencies, Means, Standard deviations, Scatter Plots, Mode	Scores of Relevance, Comparative Relevance Demographic characteristics Bar graphs of States of Residency Relevance Scores of Clinical Dietitians vs Comparative Relevance scores of Other Health Professionals and Nutritional Scientists Characterization of typical respondent
Derived Statistics Differences in Means Cronbach Alpha reliability evaluation	Unique and Shared knowledge topics Scores of Relevance and Comparative Relevance of knowledge topics for each Domain
Correlations	Correlation of Demographic characteristics to Domain Mean Relevance scores
Analysis of Variance	Prediction of relevance scores from demographic groups
Factor Analysis	Based on correlations of relevance scores of knowledge topics

Appendix C: Qualitative Interview Questions

Questions used in Qualitative Interviews
Regarding Phenomenon of Concern to Clinical Dietetics
and Terminology

1.) What do you experience as the relevant phenomena of concern in your practice of clinical dietetics?

2.) How do you think about the concept of relevance?

3.) Are any of the phenomena you mentioned more relevant than others?

4.) Do you perceive any of the phenomena of concern you discussed as unique to clinical dietetics?

5.) Do you perceive any of the phenomena of concern you discussed as being shared with other health professionals?

6.) Do you perceive any of the phenomena of concern you discussed as being shared with nutrition scientists?

7.) Should clinical dietetics expand in the future, can you imagine additional phenomena of concern that would be relevant to the practice of clinical dietetics?

8.) How would you describe the difference between a phenomenon being "relevant" and being "fundamental"?

Appendix D: Guiding Statements of The American Dietetic Association

PREAMBLE AND PRINCIPLES OF CODE OF ETHICS FOR THE PROFESSION OF DIETETICS

The American Dietetic Association and its Commission on Dietetic Registration have adopted a voluntary, enforceable code of ethics. This code, entitled the Code of Ethics for the Profession of Dietetics, challenges all members, registered dietitians, and dietetic technicians, registered, to uphold ethical principles. The enforcement process for the Code of Ethics establishes a fair system to deal with complaints about members and credentialed practitioners from peers or the public.

The first code of ethics was adopted by the House of Delegates in October 1982; enforcement began in 1985. The code applied to members of The American Dietetic Association only. A second code was adopted by the House of Delegates in October 1987 and applied to all members and Commission on Dietetic Registration credentialed practitioners. A third revision of the code was adopted by the House of Delegates on October 18, 1998, and enforced as of June 1, 1999, for all members and Commission on Dietetic Registration credentialed practitioners.

The Ethics Committee is responsible for reviewing, promoting, and enforcing the Code. The Committee also educates members, credentialed practitioners, students, and the public about the ethical principles contained in the Code. Support of the Code of Ethics by members and credentialed practitioners is vital to guiding the profession's actions and to strengthening its credibility.

PREAMBLE

The American Dietetic Association and its credentialing agency, the Commission on Dietetic Registration, believe it is in the best interest of the profession and the public it serves to have a Code of Ethics in place that provides guidance to dietetics practitioners in their professional practice and conduct. Dietetics practitioners have voluntarily adopted a Code of Ethics to reflect the values and ethical principles guiding the dietetics profession and to outline commitments and obligations of the dietetics practitioner to client, society, self, and the profession.

The Ethics Code applies in its entirety to members of The American Dietetic Association who are Registered Dietitians (RDs) or Dietetic Technicians, Registered (DTRs). Except for sections solely dealing with the credential, the Code applies to all members of The American Dietetic Association who are not RDs or DTRs. Except for aspects solely dealing with membership, the Code applies to all RDs and DTRs who are not members of The American Dietetic Association. All of the aforementioned are referred to in the Code as "dietetics practitioners." By accepting membership in The American Dietetic Association and/or accepting and maintaining Commission on Dietetic Registration credentials, members of The American Dietetic Association and Commission on Dietetic Registration credentialed dietetics practitioners agree to abide by the Code.

PRINCIPLES

1. The dietetics practitioner conducts himself/herself with honesty, integrity, and fairness.

2. The dietetics practitioner practices dietetics based on scientific principles and current information.

3. The dietetics practitioner presents substantiated information and interprets controversial information without personal bias, recognizing that legitimate differences of opinion exist.

4. The dietetics practitioner assumes responsibility and accountability for personal competence in practice, continually striving to increase professional knowledge and skills and to apply them in practice.

5. The dietetics practitioner recognizes and exercises professional judgment within the limits of his/her qualifications and collaborates with others, seeks counsel, or makes referrals as appropriate.

6. The dietetics practitioner provides sufficient information to enable clients and others to make their own informed decisions.

7. The dietetics practitioner protects confidential information and makes full disclosure about any limitations on his/her ability to guarantee full confidentiality.

8. The dietetics practitioner provides professional services with objectivity and with respect for the unique needs and values of individuals.

9. The dietetics practitioner provides professional services in a manner that is sensitive to cultural differences and does not discriminate against others on the basis of race, ethnicity, creed, religion, disability, sex, age, sexual orientation, or national origin.

10. The dietetics practitioner does not engage in sexual harassment in connection with professional practice.

11. The dietetics practitioner provides objective evaluations of performance for employees and coworkers, candidates for employment, students, professional association memberships, awards, or scholarships. The dietetics practitioner makes all reasonable effort to avoid bias in any kind of professional evaluation of others.

12. The dietetics practitioner is alert to situations that might cause a conflict of interest or have the appearance of a conflict. The dietetics practitioner provides full disclosure when a real or potential conflict of interest arises.

13. The dietetics practitioner who wishes to inform the public and colleagues of his/her services does so by using factual information. The dietetics practitioner does not advertise in a false or misleading manner.

14. The dietetics practitioner promotes or endorses products in a manner that is neither false nor misleading.

15. The dietetics practitioner permits the use of his/her name for the purpose of certifying that dietetics services have been rendered only if he/she has provided or supervised the provision of those services.

16. The dietetics practitioner accurately presents professional qualifications and credentials.

    a. The dietetics practitioner uses Commission on Dietetic Registration awarded credentials ("RD" or "Registered Dietitian"; "DTR" or "Dietetic Technician, Registered"; "CSP" or "Certified Specialist in Pediatric Nutrition"; "CSR" or "Certified Specialist in Renal Nutrition"; and "FADA" or "Fellow of The American Dietetic Association") only when the credential is current and authorized by the Commission on Dietetic Registration. The dietetics practitioner provides accurate information and complies with all requirements of the Commission on Dietetic Registration program in which he/she is seeking initial or continued credentials from the Commission on Dietetic Registration.

    b. The dietetics practitioner is subject to disciplinary action for aiding another person in violating any Commission on Dietetic Registration requirements or aiding another person in representing himself/herself as Commission on Dietetic Registration credentialed when he/she is not.

17. The dietetics practitioner withdraws from professional practice under the following circumstances:

   a.   The dietetics practitioner has engaged in any substance abuse that could affect his/her practice;

   b.   The dietetics practitioner has been adjudged by a court to be mentally incompetent;

   c.   The dietetics practitioner has an emotional or mental disability that affects his/her practice in a manner that could harm the client or others.

18. The dietetics practitioner complies with all applicable laws and regulations concerning the profession and is subject to disciplinary action under the following circumstances:

   a.   The dietetics practitioner has been convicted of a crime under the laws of the United States which is a felony or a misdemeanor, an essential element of which is dishonesty, and which is related to the practice of the profession.

   b.   The dietetics practitioner has been disciplined by a state, and at least one of the grounds for the discipline is the same or substantially equivalent to these principles.

   c.   The dietetics practitioner has committed an act of misfeasance or malfeasance which is directly related to the practice of the profession as determined by a court of competent jurisdiction, a licensing board, or an agency of a governmental body.

19. The dietetics practitioner supports and promotes high standards of professional practice. The dietetics practitioner accepts the obligation to protect clients, the public, and the profession by upholding the Code of Ethics for the Profession of Dietetics and by reporting alleged violations of the Code through the defined review process of The American Dietetic Association and its credentialing agency, the Commission on Dietetic Registration.

Mission:

Leading the future of dietetics

Vision:

American Dietetic Association members are the most valued source of food and nutrition services

Values:

Customer Focus—operates with consideration for the needs and expectations of internal and external customers

Integrity—acts ethically, with accountability and attention to excellence

Innovation—fosters an environment of positive change through creativity and continuous improvement

Life-Long Learning—takes personal accountability for own competence and seeks opportunities for continued learning

Collaboration—promotes open dialogue, cooperation and the sharing of knowledge

Inclusivity—demonstrates respect and sensitivity toward and appreciation for, the backgrounds, differences, and points of view of others

Social Responsibility—guides decisions and actions by considering economic, environmental and social implications

The American Dietetic Association Strategic Plan 2004-2008
Printed with permission

Glossary

Abductive Reasoning: making a conceptual leap from experience to arrive at an educated guess or theory about a phenomenon. Reed Adv Nurs Sci 1995; 17: 74.

Acculturation: the modification of a primitive culture by contact with an advanced culture. American Heritage Dictionary. 1979 Houghton Mifflin Dallas p. 9.

Alternative health care: often used in the sense of an either-or, instead-of, relationship to traditional medicine.

AP4: Accredited Pre-Professional Practice Program, a route of entry into the profession of dietetics.

Biomedicine: The dominant healthcare system in the US in the 20th century, including the theoretical and practice models in use by physician and other practitioners. Alternative Therapies 1997 3: p.51.

Biomedical Nutritionist: a health care giver who provides Human Biomedical Nutritional care

(see also Human Biomedical Nutrition below)

Chemopreventive agent: Nutritive or nonnutritive food component being scientifically investigated as a potential inhibitor of carcinogenesis for primary and secondary cancer prevention ... Position of The American Dietetic Association: Phytochemicals and functional foods. J Amer Diet Assoc 1995 p. 493-6.

Clinical dietetics: (theoretical definition for this work) the art of applying specialized education and training in human nutritional science and social sciences to the diagnosis and treatment of nutrition problems in clients for modifying the course of disease and/or maintaining health.

Clinical dietetics: (1917 by American Dietetic Association) the science of nutrition and the art of feeding people.

Clinical dietetics: (Kight) the emerging epidemiological branch of nutritional science.

Clinical dietitian: (1984 American Dietetic Association Role Delineation Study) a health care professional credentialed as a registered dietitian who affects the nutrition care of individuals and groups in health and illness. The clinical dietitian provides nutrition assessment, planning, implementation (including education and referral) and evaluation services; provides consultation for foodservice to coordinate nutrition care services, manages departmental and personnel functions for nutrition care services; delineates and manages external influences on the delivery of nutrition care. The clinical dietitian educates and coordinates activities as a member of the health care team; maintains skill and knowledge in optimal nutrition care; and conducts applied research.

Clinical dietitian: (operational definition for this work) refers to a clinical dietetic practitioner who qualifies for practice as demonstrated by membership in and registration with the American Dietetic Association. For purposes of this study the practitioner is also member of a selected Dietetic Practice Group.

Clinical Nutrition: describes the factors influencing intake, absorption, and metabolism of dietary factors as well as the mechanisms underlying the relation between diet and disease. Forman, MR and Halsted, CH. Am J Clin Nutr 1998; 67: 183.

Clinical practice: pertaining to a clinic or to the bedside; pertaining or founded on actual observation and treatment of patients as distinguished from theoretical or basic sciences. Dorland's Illustrated Medical Dictionary, 28th edition. 1994; WB Saunders Co, Philadelphia.

Complementary medicine: often used in the sense of together with, or complementary to, offerings of conventional medicine

Complementary and Alternative Medicine (CAM): "is a broad domain of healing resources that encompasses all health systems, modalities, and practices and their accompanying theories and beliefs, other than those intrinsic to the politically dominant health system of a particular society or culture in a given historical period. CAM includes al such practices and ideas self-defined by their users as preventing or treating illness or promoting health and well-being. Boundaries within CAM and between CAM domain and the domain of the dominant system are not always sharp or fixed". Defining and describing complementary and alternative medicine. Alternative Therapies, Panel on Definition and Description, CAM Research Methodology Conference, April 1995; March 1997 Vol. 3 No.2 p. 49-57.

CPT: Current Procedural Terminology: numerical codes used to document and report type and duration of service provided by a health care provider

CUP: Coordinated Undergraduate Program, one route of entry into the profession of dietetics.

Designer foods: "processed foods that are supplemented with food ingredients naturally rich in cancer-preventing substances" Caragay, A B. Cancer-preventive foods and ingredients. Food Tech April 1992; p. 65-8.

Designer Foods: Processed foods that are supplemented with food ingredients naturally rich in disease-preventing substances. This may involve genetic engineering of food. Position of The American Dietetic Association: Phytochemicals and functional foods. J American Diet Association 1995; p. 493-6.

Domain: a sphere of concern or function.

Drugs: "Products intended for use in the diagnosis, cure, mitigation, treatment or prevention of disease or to affect the structure or a function of the body" Federal Register June 18, 1993; 58: 33692. also Hunt JADA Feb 91 p. 151-153.

Engineered Foods: sometimes called architectured or fabricated foods, involves restructuring of food components into new entities, for example soy protein fiber into meat analogs, surimi into shrimp or crab analogs. Smith, RE. Food demands of the emerging consumer: The role of modern food technology in meeting that challenge. Am J Clin Nutr 1993 58 suppl: p. 307-12S.

Genetically Engineered Foods: the insertion of one or more genes with a clearly defined and desired function into plants, animals or microorganisms that are used for human food or animal feed; this process can make use of any genes regardless of their original source, genes do not necessarily have to be from closely related plants or animals that inter-breed. Brody, J E. A cool look at genetically altered foods. Food Insight IFIC Foundation 1993; July/August p. 2-3.

Foods: "products primarily consumed for their taste, aroma, or nutritive value". Food and Drug Administration. Regulation of Dietary Supplements. Federal Register June 18, 1993; 58: 33692. also Hunt JADA Feb 91; p. 151-153.

Functional foods: "processed foods, defined by the main functional ingredient; (oligosaccharides, fibres, minerals, etc.), that are claimed to perform specific health roles such as preventing, treating and curing various diseases, and which fall into a category somewhere between food, dietary supplements and drugs". Griffen, G. Talking straight: the benefits to industry of communicating advances in food sciences and diet philosophy. Trends Food Sci Technol 1993 4: p.77-9. also Hunt JADA Feb 91; p.151-153.

Functional Food: Any modified food or food ingredient that may provide a health benefit beyond the traditional nutrients it contains. Position of The American Dietetic Association: Phytochemicals and functional foods. J American Diet Association; 1995 p. 493-6.

HMO: Health Maintenance Organization, a business concerned with the contracting and provision of health care to a defined population.

Human Biomedical Nutrition: specialized clinical caregiving based on the interactions between advanced-level, primarily nutriologic-focused practitioner, the 3P Standards Minimum Data Set Structured 9-step Nutritional Care process and her/his caseload of medically classified individuals.

Medical Foods: "products intended for use under the supervision for specific dietary management of a disease or condition". Food and Drug Administration. Food labeling: mandatory status of nutrition labeling and nutrient content revision, format for a nutrition label. Federal Register January 6, 1993; 58: 2151. also Hunt JADA Feb 91 p. 151-153.

Medical foods definition has evolved
> from meaning enteral formulas,

>> to "foods represented for use solely under medical supervision to meet nutritional requirements in specific conditions" (1973);

>> to the first legal definition in 1988 "foods formulated for consumption or administration enterally under the supervision of a physician and intended for the spe-

cific dietary management of diseases or conditions for which distinctive nutritional requirements based on recognized scientific principles are established by 'medical evaluation'". Scientific Status Summary, Institute of Food Technologists' Expert Panel on food safety and nutrition. Medical foods. Food Tech: April, 1992: 8796.

Medical Nutrition Therapy: Medical nutrition therapy involves the assessment of the nutritional status of patients with a condition, illness or injury that puts them at risk. This includes review and analysis of medical history, laboratory values and anthropometric measurements. Based on the assessment, nutrition modalities most appropriate to manage the condition or treat the illness or injury are chosen and include the following:

Diet modification and counseling leading to the development of a personal diet plan to achieve nutritional goals and desired health outcomes.

Specialized nutrition therapies including supplementation with medical foods for those unable to obtain adequate nutrients through food intake only; enteral nutrition delivered via tube feeding into the gastrointestinal tract for those unable to ingest or digest food; and parenteral nutrition delivered via intravenous infusion for those unable to absorb nutrients. Position of The American Dietetic Association: Cost-effectiveness of medical nutrition therapy J American Diet Association Jan 1995 p. 89.

Medicine: Strict constructionist definition: biomedicine and physician practitioners.

Broad definition: everything that people do to preserve, promote and restore health and to prevent and treat illness

Metaparadigm: a statement of global, abstract over-reaching concepts that describes the domains, or phenomena of concern to a profession. It represents the actual and/or potential reality for a discipline and acts as a framework that encompasses the less abstract levels of knowledge in the structural hierarchy of the body of knowledge.

The proposed Metaparadigm of Clinical Dietetics includes the domains of Reference Person, Human Condition, Practitioner Actions/Attitudes, Practitioner Environment, Client Actions/ Attitudes, Client Environment and Nutraceuticals.

Metascope: a diagram for viewing the perceptual range of abstraction and associations encompassing the professional realms validated.

A metascope illustrates the links between the initial primary domains of the Metaparadigm of Clinical Dietetics, the Clinical Dietetic Factors and the expanded (metascopic) labels derived from the Eight World Hypotheses. The essence of a clinical dietetic theorist's work can be expressed in a metascope, demonstrating the conceptual range of her concerns.

Metascopic Label: Expanded (metascopic) labels, induced by factor analysis from practitioners perceptions of knowledge topics and linked to the Eight World Hypotheses. Metascopic labels include: Knowledge Base, World Views, Contextual Links, Organismic Components, Psychosocial Nutrology, Improvement Knowledge and Integrated Knowledge.

Noumenon: an object of purely intellectual intuition, as opposed to an object of sensuous perception The American Heritage Dictionary New College Edition 1979.

Nutraceutical (source definition): "any substance that maybe considered a food or part of a food and provides medical or health benefits, including the prevention and treatment of disease. Nutraceuticals may range from isolated nutrients, dietary supplements, and diets to genetically engineered "designer" foods, herbal products, and processed products, such as cereals, soups and beverages. The Nutraceutical Initiative: A proposal for economic and regulatory reform. Food Technology. ed. Donald Pszczola. April, 1992 p. 77-78.

Nutraceutical: Any substance that may be considered a food or part of a food and provides medical or health benefits, including the prevention and treatment of disease … Position of The American Dietetic Association: Phytochemicals and functional foods. J American Diet Association 1995 p. 493-6.

Nutritional Epidemiology: examines the existence of, the determinants of, and the magnitude of disease risk in relation to diet; biochemical, anthropometric, and clinical indicators of nutritional status; as well as the public health effect of this risk on the population. Forman, MR and Halsted, CH. Am J Clin Nutr 1998; 67: 183.

Nutritional Scientist: refers to academic scientists who elucidate problems and solutions related to nutritional status, nutritional requirements, nutrient metabolism and/or biochemistry, molecular and genetic biology related to nutrition.

Other Health Professional: refers to any other health profession/professional about which respondent has knowledge or experience.

Pharmafood: Food or nutrient that claims medical or health benefits, including the prevention and treatment of disease. Position of The American Dietetic Association: Phytochemicals and functional foods … J American Diet Association 1995 p. 493-6.

Phytochemicals: every naturally occurring chemical substance present in plants; some are biologically active, some are not. Caragay, A B. Cancer-preventive foods and ingredients. Food Tech April 1992; p. 65-8.

Phytochemical: "Substances found in edible fruits and vegetables that may be ingested by humans daily in gram quantities and that exhibit a potential for modulating human metabolism in a manner favorable for cancer prevention." Position of The American Dietetic Association: Phytochemicals and functional foods. J American Diet Association; 1995 p. 493-6.

Phytonutrients: unique substances produced during the natural course of plant growth and development that are specific to each plant's genes and environment. Bland, J S. Phytonutrition, phytotherapy, and phytopharmacology. Alt Ther Nov. 1996 2 p. 73-6.

Phytonutrition: the role of these substances in cultural food practices and cuisines worldwide in supporting health. Bland, J S. Phytonutrition, phytotherapy, and phytopharmacology. Alt Ther 1996 2 p. 73-6.

Phytotherapeutic role: phytonutrients acting as modifiers of physiological function Bland, J S. Phytonutrition, phytotherapy, and phytopharmacology. Alt Ther 1996; 2: p. 73-6.

Relevance: refers to any phenomenon of concern that pertains to, is influential, important or applicable to the practice of clinical dietetics, other health professions or to nutritional science or is an integral part of the body of knowledge of each, respectively.

Shared: To participate in, use, or experience in common, can be applied to intangible things; in this study: knowledge (American Heritage Dictionary of the English Language, New College Edition, 1979)

WIC: Women, Infants and Children, a federally funded program to provide health care and nutritional care to defined populations.

Reference List
(In Alphabetical Order)

American Dietetic Association. New code of ethics developed for the dietetic profession. Amer Diet Assoc Courier; 1988; 27 (6):1-2.

American Dietetic Association. Role Delineation's for Advanced-Level and Specialty Practice in Dietetics: Results of an Empirical Study. Technical Report. Macmillan/McGraw-Hill, Monterey, CA. 1992;1: Chapter 5, p. 5-25.

(Position of the) American Dietetic Association: Phytochemicals and functional foods. J Amer Diet Assoc 1995; 95: 493-6.

Baird SC, Burrelli J, Flack H. Role delineation and verification for entry-level positions in clinical dietetics. Chicago, IL: American Dietetic Association. 1984.

Bierman EL. President's Address, 1992: The American Society for Clinical Nutrition-a glance backward, a look ahead. Amer J Clin Nutr 56: 726-729.

Biesecker, RL. Clinical Nutrition Expert Status as Related to Selected Demographic, Diagnostic Thinking, Knowledge and Motivational Variables. Unpublished dissertation. 1999. The University of Arizona.

Blackburn GL. Presidential address: interaction of the science of nutrition and the science of medicine. J Parenteral and Enteral Nutr 1979; 3: 131-135.

Blegan, MA, Tripp-Reimer T. Implications of nursing taxonomies for middle-range theory development. Adv Nurs Sci 1997; 19: 37-49.

Brownell KD, Cohen LR. Adherence to dietary regimens 2: components of effective interventions. Behav Med 1995; 20: 155-164.

Burns, N, Grove, SK. The Practice of Nursing Research. 2nd ed. Philadelphia: W.B. Saunders, 1993 p.382.

Carey M. Diabetes guidelines, outcomes, and cost-effectiveness study: a protocol, prototype, and paradigm. J Amer Diet Assoc 1995; 95: 976-78.

(The) Chicago Dietetic Association and The South Suburban Dietetic Association. Nutrition management of the patient with psychiatric disorders. In: Manual of clinical dietetics fifth edition. Chicago: American Dietetic Association: 1996: 685-713.

Christie BW, & Kight MA. Educational empowerment of the clinical dietitian: A proposed practice doctorate curriculum. J Amer Diet Assoc 1993; 2: 173-176.

Codes of Federal Regulations, Title 29, Part 541, Section 302. Washington DC: Office of the Federal Register, National Archives and Records Administration; 1991, p. 336.

Cody W K, Mitchell G J. Parse's theory as a model for practice: the cutting edge. Adv Nurs Sci 1992 15: 52-65.

Committee on Clinical Practice Issues in Health and Disease. The role and identity of physician nutrition specialists in medical-school affiliated hospitals. Amer J Clin Nutr 1995; 61: 264-268

Cooper LF. Florence Nightingale's contribution to dietetics 1954. In: Beeuwkes AM, Todhunter EN, Weigley ES.(Eds) Essays on history of nutrition and dietetics. Chicago, IL: American Dietetic Association, 1967: 5-11.

Coulston AM, Rock CL. A summary of the current state of knowledge in clinical nutrition and dietetic practice: suggestions for future research in dietetic practice and implications for health care. Proceedings of the Research Agenda for Dietetics Conference; May 14-15, 1992; Chicago. Chicago: American Dietetic Association. 1992: 1-24.

Cousins R. Incoming president's column: nutrition notes. American Society for Nutritional Sciences 1996; 32 (2): 1-2.

Covey S. The Seven Habits of Highly Effective People. New York: Simon and Schuster, 1989: 24-27.

Davis, CA, Britten P, Myers, EF. Past, present and future of the Food Guide Pyramid. J Amer Diet Assoc. 2001 (8) 881-885.

DeBusk R. Genetics and Nutrition: The future is now. Nutrition in Complementary Care, a Practice Group of The American Dietetic Association; 2000. 1-12. <http://complementarynutri-tion.org>; article ID=AR00002.

Diagnostic and Statistical Manual of Mental Disorders, fourth edition. Washington, DC: American Psychiatric Association, 1994: 251-254.

Dickoff J, James P. A theory of theories: A position paper. Nurs Res 1968; 17: (3): 197-203.

Dillman DA. Mail and Telephone Surveys. New York JohnWiley & Sons, Inc. 1978.

Dubin R. Theory building. New York: The Free Press, 1978: 5-7.

Dubos R. The intellectual basis of nutrition science and practice. Nutrition Today 1979, July/August: 31-34.

Dutra-de-Oliveira JE, Marchini JS. Clinical nutrition for MD's: reappraisal and identity (letter to the editor) American Journal of Clinical Nutrition 1995; 62 ; 1289-1290.

Dwyer, JT. Scientific underpinnings for the profession: Dietitians in research. J Amer Diet Assoc 1997; 97: 593-597.

Eckberg DL, Hill, L Jr. The paradigm concept and sociology: a critical review. Amer Sociol Rev 1979; 44: 925-37.

Ericsson KA, Krampe, RT, Tesch-Romer, C. The role of deliberate practice in the acquisition of expert performance. Psychological Review 1993; 100: 363-405.

Fawcett J. The metaparadigm of nursing : present status and future refinements. Image: J Nurs Scholarship 1984; 16: 84-7.

Fawcett J. Analysis and evaluation of conceptual models in nursing. Philadelphia: F.A.Davis, 1995.

Fink III, JL, Counseling patients on Rx therapy and OTC drugs: legal issues. The Pharmacy Letter Student Edition 1997; 1: 1-4.

Fitz P. President's page: Show them the numbers. J Amer Diet Assoc 1997; 97: 898.

Food and Nutrition Board Concept Paper. How should the Recommended Dietary Allowances be revised? Nutrition Reviews 1994; 52: 218.

Forcier JI, Kight MA, Sheehan ET. Point of view: acculturation in clinical dietetics. J Amer Diet Assoc. 1977; 70: 181-185.

Forman MR, Halsted CH. Nutritional epidemiology in The American Journal of Clinical Nutrition. Am J Clin Nutr 1998; 67: 183

Frank-Spohrer GC. Community nutrition. Gaithersburg, MD: Aspen Publishers, Inc. 1996: 84-107.

Gerstle, JF, Varenne H, Contento I. Post-diagnosis family adaptation influences glycemic control in women with type 2 diabetes. J Amer Diet Assoc. 2001; (8) 918-992.

Halsted CH. Certification of clinical nutritionists. Amer J Clin Nutr 1995; 62: 10-12

Harper, A. The science and the practice of nutrition: reflections and directions. Am J Clin Nutr. 1991; 53: 413-20.

Healthy People 2000. Washington, D C: U.S. Department of Health and Human Services, Public Health Service 1991 1: 112-113.

Huls, A.. Nutrition Parameters predicting functional status decline in the older adult. UMI Dissertation Services, Ann Arbor Michigan. 1999.

International Classification of Diseases, 9th Revision. Geneva Switzerland: World Health Organization 1977.

Insull Jr. W. Dietitians as intervention specialists: a continuing challenge for the 1990s. J Amer Diet Assoc 1994; 94: 551-2.

_____. Table of Contents. J Amer Diet Assoc. 2001; 6.

Kein KS, Johnson CA, Gates GE. Learning needs and continuing professional education activities of *Professional Development Portfolio* participants. J Amer Diet Assoc. 2001; 6: 697-702.

Kight, MA, Nutritional Caregiving Position Statement. Diagnostic Nutrition Network. 2001; 10 (1) 1 (April, 2001)

Kight MA. A dietetic-specific diagnostic reasoning approach to communicating in code. Diagnostic Nutrition Network 1993; 2 (2): 2-3.

Kight MA. Emerging concept development. Diagnostic Nutrition Network 1995; 4 (2): 8.

Kight MA. Perceptions of a research typology for client-centered dietetics. Dietitians in Critical Care 1986; 8: 2.

Kline P. An Easy Guide to Factor Analysis. New York: Routledge 1994: 64-5.

Koop CE. The surgeon general's report on nutrition and health. Bethesda, (MD): U.S. Department of Health and Human Services; 1988 DHHS Publication No. 88-50210.

Leyse, RL, Perceptions of the Metaparadigm of Clinical Dietetics. UMI Dissertation Services, Ann Arbor Michigan. 1998. pages 83-84.

Leyse RL, Kight MA. A metaparadigm of clinical dietetics: the template for building our unique body of knowledge. Diagnostic Nutr Network 1993; 4: 2-5.

MacEachern MT. The hospital dietary department-a forecast. J Amer Diet Assoc. 1925; (1): 3-8.

Mason JB, Habicht JP, Greaves JP, Jonsson U, Kevany J, Martorell R, Rogers B. Public nutrition. (Letter to the Editor). Amer J of Clin Nutr 1996; 63: 399-400.

Mason M, Winslow R. Professional roles of dietitians: do dietitians and physicians agree? Nutr Rev 1994; 52: (9) 315-317.

McClosky JC, Bulechek GM. Nursing Interventions Classification St. Louis, MO: Mosby Year Book. 1992.

McLaren DS. Nutrition in medical schools: A case of mistaken identity. Amer J Clin Nutr 1994; 59: 960-3.

Meleis AI. Hypatia, Hatshepsut, and nursing scholars: scholars in disguise. In: Theoretical Nursing: Development and Progress. New York: J. B. Lippincott, 1991: 115-125.

Monsen ER, Cheney CL. Research methods in nutrition and dietetics: design, data analysis and presentation. J Amer Diet Assoc. 1988; 88: 1047-1065.

Muller ME, Dzurec LC. The power of the name. Adv Nurs Sci 1993; 15: 15-22.

National Nutrition Monitoring and Related Research Act of 1990. United States Congress. House. 101st Congress. [Reports: no. 101-788 (Committee on Agriculture), Congressional Record, Vol. 136] 1990.

National Research Council. Recommended Dietary Allowances, tenth ed. Washington, D.C. National Academy Press; 1989.

North American Nursing Diagnosis Association. Taxonomy I-Revised 1990. St. Louis: The North American Nursing Diagnosis Association.

Nunnally JC, Bernstein IH. Psychometric Theory. San Francisco: McGraw-Hill Inc. 3rd ed. 1994 p. 317.

The Nutraceutical initiative: A proposal for economic and regulatory reform. In: Pszcola, D. Ed. Food Technology. April, 1992; 77-8.

Olson R. Wendall H. Griffith (1985-1968): a biographical sketch. J Nutr 1986; 116 (12): 2327-2339.

Panel on Definition and Description. Defining and describing complementary and alternative medicine. CAM research methodology conference; April 1995. Alt Ther 1997; 3: 49-57.

Parks S. Challenging the future—an evolving global perspective for the profession. J Amer Diet Assoc 1994; 94: 782-84.

Pepper SC. World hypotheses: A study in evidence. Cambridge, England: Cambridge University Press. 1942. Reprinted California, University of California Press, 1961.

Pew Health Professions Commission. Critical challenges: revitalizing the health professions for the twenty-first century. San Francisco CA: University of California at San Francisco, Center for the Health Professions; 1995.

Pew Health Professions Commission. Health professions education for the future: Schools in service to the nation. San Francisco CA: University of California at San Francisco, Center for the Health Professions; 1993.

Pew Health Professions Commission. Healthy America: Practitioners for 2005. San Francisco CA: University of California at San Francisco, Center for the Health Professions; 1991.

Pincus, T. Analyzing long-term outcomes of clinical care without randomized controlled clinical trials: the consecutive patient questionnaire database. Advances 1997; 13: 3-32

Reed PG, A treatise on nursing knowledge development for the 21st century: beyond postmodernism. Adv Nurs Sci 1995; 17: 70-84.

Romon M, Nuttens MC, Vambergue A, Verier-Mine O, Biausque S, Lemaire C, Fontaine P, Salomez JL, Beuscart R. Higher carbohydrate intake is associated with decreased incidence of newborn macrosomia in women with gestational diabetes. J Amer Diet Assoc; 2001: (8) 897-902.

Rothman KJ, Michels KB. The continuing unethical issue of placebo controls. N Eng J Med 1994; 331: 394-7.

Rubenstein RA, Laughlin Jr., CD, McManus J. Science as cognitive process. Philadelphia: University of Pennsylvania Press; 1984.

Rubik, B. 1996) Personal Communication. Why is Alternative Medicine Alternative? Department of Integrative Medicine, University of Arizona School of Medicine, April 4, 1996.

Salant P, Dillman DA. How to Conduct Your Own Survey. John Wiley & Sons, Inc.. New York. 1994.

Schuftan, C. Technical, ethical and ideological responsibilities in nutrition. World Review Nutrition and Dietetics. 1987; 53: 1-27.

Schwartz, GE, Russek, L.G. The challenge of one medicine: Theories of health and eight "World Hypotheses". Advances, J Mind-Body Health 1997; 13: 7-23.

Simopoulos AP, Genetic variation and nutrition. Nutr Today 1995; 30: 194-206.

Smith MC. Proposed metaparadigm for nursing research and theory development. Image: J Nurs Scholarship 1979; 11: 75-9.

Splett, P. and Myers, EF, A proposed model for effective nutrition care, JADA, Volume 101 No. 3, p. 357-363.

Thomas PR, Earl R. Creating the future of the nutrition and food sciences. J Amer Diet Assoc 1994: 257-9.

Thomson CA, Kight MA. Influence techniques and activities clinical dietitians use when interacting with physicians. J Amer Diet Assoc 1990; 9: 80-84.

Todhunter EN. The evolution of nutrition concepts—perspectives and new horizons. 1964. In: Beeuwkes AM, Todhunter EN, Weigley ES, Editors. Essays on History of Nutrition and Dietetics. Chicago, IL: The American Dietetic Association. 1967: 12-20.

Tripp-Reimer T, Woodworth G, McClosky JC, Bulechek G. The dimensional structure of nursing interventions. Nurs Res 1996; 45: 10-17.

Walker LO, Avant KC. Strategies for theory construction in nursing. San Mateo, CA: Appleton and Lange, 1988: 22.

Walker RS. Scientists as advocates in the policy process. Federation of American Societies for Experimental Biology 1995; 28 (9): 2-3.

Walsh R, The spirit of evolution. Noetic Sciences Review 1995: 17-41.

Wertheimer AI, Smith, MC. Pharmacy practice: social and behavioral aspects. John P. butler, ed. Third edition. Baltimore: Williams and Wilkins; 1989.

Wheatley, Margaret J. Leadership and the New Science: Discovering Order in a Chaotic World. 2nd ed. Berrett-Koehler Publ. San Francisco; 1999.

Young, VR & Scrimshaw, NS. Genetic and Biological Variability in Human Nutrient Requirements. Figure by Arroyave, G. Am J Clin Nutr 1979; 32: 488.

Young EA. The physician-dietitian roles in advancing medical nutrition therapy. Perspectives Appl Nutr 1995; 2 (3): 25-26.

Dance of Clinical Dietetics

Dance with a theory and dance with some fat,
Dance with a mineral or dance with a rat,
Dance with some fiber or dance with caffeine,
Dance with a calorie that cannot be seen.

Chorus: And we'll dance, dance, dance;
        Come dancing with me, aha;
        And we'll dance, dance, dance
        Come dancing with me.

Dance with a vitamin, dance with your weight,
Dance with instruction to clean up your plate,
Dance with your snack foods and dance with your meals.
Pay close attention to how your food feels.
Chorus

Dance with anemia, dance with some starch,
Join the fundraising and dance with a march.
Dance with the science and dance with the art,
Watch your cholesterol and dance from your heart.
Chorus

Dance with each PERSON, all stages and types,
Dance with a REFERENCE to cut out the hype.
Dance with CONDITIONS of all HUMAN kind,
Dance with the body and dance with the mind.
Chorus

Dance with concern over food faddists' claims,
All NUTRACEUTICALS are our domain.
Dance with the public and teach them the score.
Dance with our knowledge, we'll always need more.
Chorus

Dance with PRACTITIONERS—ACTIONS they take,
Dance with their ATTITUDES—make no mistake.
Dance our profession and all it entails,
Our Image, our standards; you know we're not frail.
Chorus

Dance with our CLIENTS, especially their ACTS,
Change their ENVIRONMENT to help them relax.
Dance with philosophy; know what you think,
Dance your beliefs and dance close to the brink.

"Dance of Clinical Dietetics" by Ruth Leyse, MS, RD, LD October, 1992
Inspired by original music and lyrics of "Dance" by Jim Manley

The "Metaparadigm of Clinical Dietetics" dissertation completed and the
PhD degree awarded to Dr. Leyse in March, 1998

Survey Instrument for
The Metaparadigm of Clinical Dietetics

THE UNIVERSITY OF
ARIZONA.
TUCSON ARIZONA

November, 1996

Dear Fellow Clinical Dietitian,

Thank you for committing approximately one hour of your time to participate in the Metaparadigm of Clinical Dietetics Validation Survey. Together we can help clinical dietetics gain recognition by defining the phenomena of concern to our profession and by comparing our relevant body of knowledge to that of other health professionals and nutritional scientists.

Returning your completed survey will constitute informed consent for me to analyze and publish summary results from analysis of the data. Confidentiality will be maintained: responses will be coded for analysis. No individual's responses will be considered or reported separately.

Enclosed are:
- an overview of the survey
- the instruction sheet
- the definition sheet
- the survey, printed on the computer-read answer sheet
- a postage-paid, addressed manila envelope for returning the survey

If you have any questions, feel free to call me at (520) 323-7168 between 6:30 AM and 7:30 AM MST time or at e-mail address RUTHLEYSE@AOL.com.

I will be looking forward to receiving your completed survey in the next two-three weeks.

Sincerely,

Ruth Leyse, MS, RD.

Ruth Leyse, MS, RD
Doctoral Candidate

encl: 5

Metaparadigm of Clinical Dietetics Validation Survey

Survey Overview

The purpose of this survey is to determine the **perceptions of Clinical Dietitians** regarding the **relevance of selected knowledge topics** drawn from the body of knowledge of clinical dietetics and/or that of other health professionals or nutritional scientists.

Part I. examines the **Relevance** of each knowledge topic to **Clinical Dietetics**.

Part II. examines the **Comparative Relevance** of each knowledge topic **to Other Health Professionals**.

Part III. examines the **Comparative Relevance** of each knowledge topic **to Nutritional Scientists**.

The abstract concepts used to describe the most general level of the body of knowledge relevant to a professional group are collectively called a **Metaparadigm**. It gives direction to the development of paradigms, theories and science of the profession, as illustrated below:

Concepts proposed for the Metaparadigm of Clinical Dietetics are: Reference Person, Human Condition, Practitioner Actions and Attitudes, Practitioner Environment, Client Actions & Attitudes, Client Environment, and Nutraceuticals. This survey groups selected knowledge topics into these seven domains and requests your perceptions of their relevance to clinical dietitians and others.

Metaparadigm of Clinical Dietetics						
Reference Person	Human Condition	Practitioner Actions / Attitudes	Practitioner Environment	Client Actions/ Attitudes	Client Environment	Nutraceuticals

The domains, knowledge topics and spaces to express your perceptions start on page four.

Metaparadigm of Clinical Dietetics Validation Survey

Instruction Sheet

The survey and the answer sheet have been combined for ease in recording your perceptions and scanning your responses by computer.

Use a number 2 pencil to mark your answers.

Mark only one answer to most items.
A few items request that you mark "any that apply". On these items, mark as many as accurately reflect your thoughts.

On the first page of the survey, in the upper left corner in the shaded box please write in the two-letter postal service abbreviation of the state in which you practice and the number of your primary Dietetic Practice Group (primary = 50% or more of your employment time).

Dietetic Practice Group numbers:

10 - Public Health Nutrition	24 - Dietitians in Nutrition Support
11 - Gerontological Nutritionists	25 - Dietitians in Physical Medicine and
12 - Dietetics in Developmental	Rehabilitation
and Psychiatric Disorders	27 - Dietitians in General Clinical Practice
20 - Oncology Nutrition	28 - Perinatal Nutrition
21 - Renal Dietitians	31 - Consultant dietitians in Health Care
22 - Pediatric Dietitians	Facilities
23 - Diabetes Care and Education	33 - Sports, Cardiovascular and Wellness
	Nutritionists

You may complete the survey at your own pace. It does not have to be completed in one sitting.

You may first complete all of "Part I - Clinical Dietetics", then "Part 2 - Other Health Professionals" and then proceed to "Part 3 - Nutritional Scientists" if you like. Or you may record your perceptions concerning all three groups for each individual knowledge topic as you proceed through the survey.

For your convenience the sheet of definitions is unattached. It provides the survey definitions for major terminology and the abbreviations for the response categories. You may want to keep it handy for reference during completion of the survey. Metaparadigm Domain definitions are included in the body of the survey where they are being considered.

Definitions:

<u>Clinical Dietitian</u> refers to a member of the American Dietetic Association who has passed the registration examination and has specified a clinical area of interest on membership data forms.

<u>Other Health Care Professional</u> refers to **any** other health profession/professional about which you have knowledge or experience. (You will be asked to designate which other health professions/professionals you are considering for each domain.)

<u>Nutritional Scientist</u> refers to academic scientists who elucidate problems and solutions related to nutritional status, nutritional requirements, nutrient metabolism and/or biochemistry, molecular and genetic biology related to nutrition.

<u>Relevance</u> refers to any phenomenon of concern that pertains to, is influential, important or applicable to the practice of clinical dietetics, other health professions or to nutritional science or is an integral part of the body of knowledge of each, respectively.

<u>Domain</u> refers to a sphere of concern or function.

Instructions:

<u>Part I</u> : **Indicate** the term that best describes **your perception of Relevance** of all items to Clinical Dietetics.

Use the following scale to indicate your perceptions of **Relevance to Clinical Dietetics.**

> IR = **Irrelevant**
> SIR = **Somewhat Irrelevant**
> SR = **Somewhat Relevant**
> R = **Strongly Relevant**

<u>Part II:</u> **Indicate** the term that best describes **your perception of Comparative Relevance** of all items to **Other Health Professionals.**

Use the following scale to indicate the **Comparative Relevance** to Other Health Professionals.

> NR = **Not Relevant**
> LR = **Less Relevant than to Clinical Dietetics**
> ER = **Equally as relevant as to Clinical Dietetics**
> MR = **More Relevant than to Clinical Dietetics**

<u>Part III:</u> **Indicate** the term that best describes **your perception of Comparative Relevance** of all items to **Nutritional Scientists.**

Use the **same scale of Comparative Relevance** for Nutrition Scientists as you used for Other Health Professionals

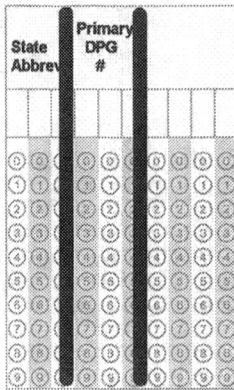

State Abbrev | Primary DPG #

A. Domain of Reference Person

Reference Person refers to the theoretical, statistically derived individual representative of the reference population, for example the reference "infant 0.5-1.0 years old" referred to in the Recommended Dietary Allowances. (National Research Council, 1989). It includes the assumption of defined criteria of selection and assumes the user is informed regarding the essential details of the derivation. When reference values are used in evaluation or interpretation, it is acknowledged that health and disease are relative, not absolute states.

Knowledge Topics	Part I Clinical Dietetics	Part II Other Health Professionals	Part III Nutritional Scientists

A-1. Normal Human Functioning

	Part I	Part II	Part III
1a. Nutrient disposition (absorption, utilization, excretion)	IR SIR SR R	NR LR ER MR	NR LR ER MR
1b. Nutrient functions	IR SIR SR R	NR LR ER MR	NR LR ER MR
1c. Nutritional factors in hormone regulation & gene expression	IR SIR SR R	NR LR ER MR	NR LR ER MR
1d. Integrated metabolism	IR SIR SR R	NR LR ER MR	NR LR ER MR
1e. Normal appearance of tissues likely to develop nutrient-based lesions	IR SIR SR R	NR LR ER MR	NR LR ER MR

A-2. Normal Human Developmental Stages

	Part I	Part II	Part III
2a. Aging	IR SIR SR R	NR LR ER MR	NR LR ER MR
2b. Growth patterns	IR SIR SR R	NR LR ER MR	NR LR ER MR
2c. Normal reproduction	IR SIR SR R	NR LR ER MR	NR LR ER MR

A. Domain of Reference Person ,continued

Knowledge Topics	Part I Clinical Dietetics	Part II Other Health Professionals	Part III Nutritional Scientists

A-3. Desired Nutritional Health Status

	Part I	Part II	Part III
3a. Acceptable laboratory test ranges	IR SIR SR R	NR LR ER MR	NR LR ER MR
3b. Recommended Dietary Allowances	IR SIR SR R	NR LR ER MR	NR LR ER MR

A-4. Scientific Methodology for Nutrition and Dietetics

	Part I	Part II	Part III
4a. Anthropometric measurement	IR SIR SR R	NR LR ER MR	NR LR ER MR
4b. Nutritional science laboratory methods	IR SIR SR R	NR LR ER MR	NR LR ER MR
4c. Nutritional reference values	IR SIR SR R	NR LR ER MR	NR LR ER MR
4d. Statistical methods	IR SIR SR R	NR LR ER MR	NR LR ER MR
4e. Theory-building in clinical dietetics	IR SIR SR R	NR LR ER MR	NR LR ER MR

Which "other health professionals" were you considering when you completed this domain? Indicate any that apply.

Nurses ☐   Pharmacists ☐   Physicians ☐   Psychologists ☐   Other (please specify_____) ☐

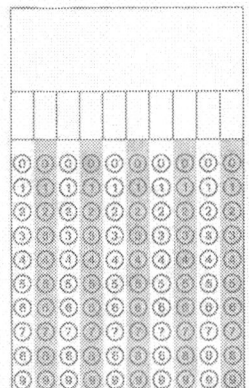

B. Domain of Human Condition

For clinical dietitians Human Condition refers to the nutritional status of individuals in a state of health or with nutritional problems. The scientifically derived reference status is compared with observed departures from "Normal" status to assess the human condition of clients. Such assessment gives direction to practitioner actions related to clients.

| Knowledge Topics | Part I
Clinical Dietetics | | | | Part II
Other Health Professionals | | | | Part III
Nutritional Scientists | | | |
|---|---|---|---|---|---|---|---|---|---|---|---|---|

B-1. Pre-Clinical Sciences

	Part I Clinical Dietetics				Part II Other Health Professionals				Part III Nutritional Scientists			
1a. Anthropology	IR	SIR	SR	R	NR	LR	ER	MR	NR	LR	ER	MR
1b. Biochemistry	IR	SIR	SR	R	NR	LR	ER	MR	NR	LR	ER	MR
1c. Biology	IR	SIR	SR	R	NR	LR	ER	MR	NR	LR	ER	MR
1d. Physiology	IR	SIR	SR	R	NR	LR	ER	MR	NR	LR	ER	MR
1e. Psychology	IR	SIR	SR	R	NR	LR	ER	MR	NR	LR	ER	MR
1f. Epidemiology	IR	SIR	SR	R	NR	LR	ER	MR	NR	LR	ER	MR

B-2. Knowledge Concerning Individuals' Departure From Normal Human Functioning

	Part I Clinical Dietetics				Part II Other Health Professionals				Part III Nutritional Scientists			
2a. Changes in tissue appearance related to nutritional deficiency	IR	SIR	SR	R	NR	LR	ER	MR	NR	LR	ER	MR
2b. Individual's food allergies/sensitivities	IR	SIR	SR	R	NR	LR	ER	MR	NR	LR	ER	MR
2c. Signs/symptoms/ potential for disease development related to nutritional status/ function	IR	SIR	SR	R	NR	LR	ER	MR	NR	LR	ER	MR

B. Domain of Human Condition, continued

Knowledge Topics	Part I Clinical Dietetics	Part II Other Health Professionals	Part III Nutritional Scientists

B-3. Nutritional Status

	Part I	Part II	Part III
3a. Expression of findings using nutritional diagnostic codes, classification systems	IR SIR SR R	NR LR ER MR	NR LR ER MR
3b. Interaction of lifestyle status, health status, with nutritional status	IR SIR SR R	NR LR ER MR	NR LR ER MR
3c. Prognostic Nutrition Indices	IR SIR SR R	NR LR ER MR	NR LR ER MR
3d. Potential or presence of nutrient/food/drug interactions	IR SIR SR R	NR LR ER MR	NR LR ER MR

Which "other health professionals" were you considering when you completed this domain? Indicate any that apply.

Nurses ☐   Pharmacists ☐   Physicians ☐   Psychologists ☐   Other (please specify_____) ☐

C. Domain of Practitioner Actions / Attitudes

Practitioner refers to Clinical Dietitian.
Actions refer to behaviors engaged in or purposefully refrained from, relative to the practice of the profession or to professional development.
Attitude refers to intrapersonally based thoughts and feelings about aspects of the clinical dietitian's professional role enactment that are elicited by situational cues. They may be explicit and willfully affect professional behavior or implicit (unstated, unknown, subconscious or unconscious) and involuntarily affect behavior, not being under the influence of the will. In this instrument attitude includes, but is not limited to, ability, aptitude, beliefs, decisions, emotions, ethics, ideas, knowledge, morals, opinions, preferences, thoughts, values, will and world view.

C-1. Assessment of Individuals

	Part I	Part II	Part III
1a. Diet (eating habits, food preferences)	IR SIR SR R	NR LR ER MR	NR LR ER MR
1b. Environmental (cultural influences, food availability, food preparation facilities, social support)	IR SIR SR R	NR LR ER MR	NR LR ER MR

C. Domain of Practitioner Actions / Attitudes, continued

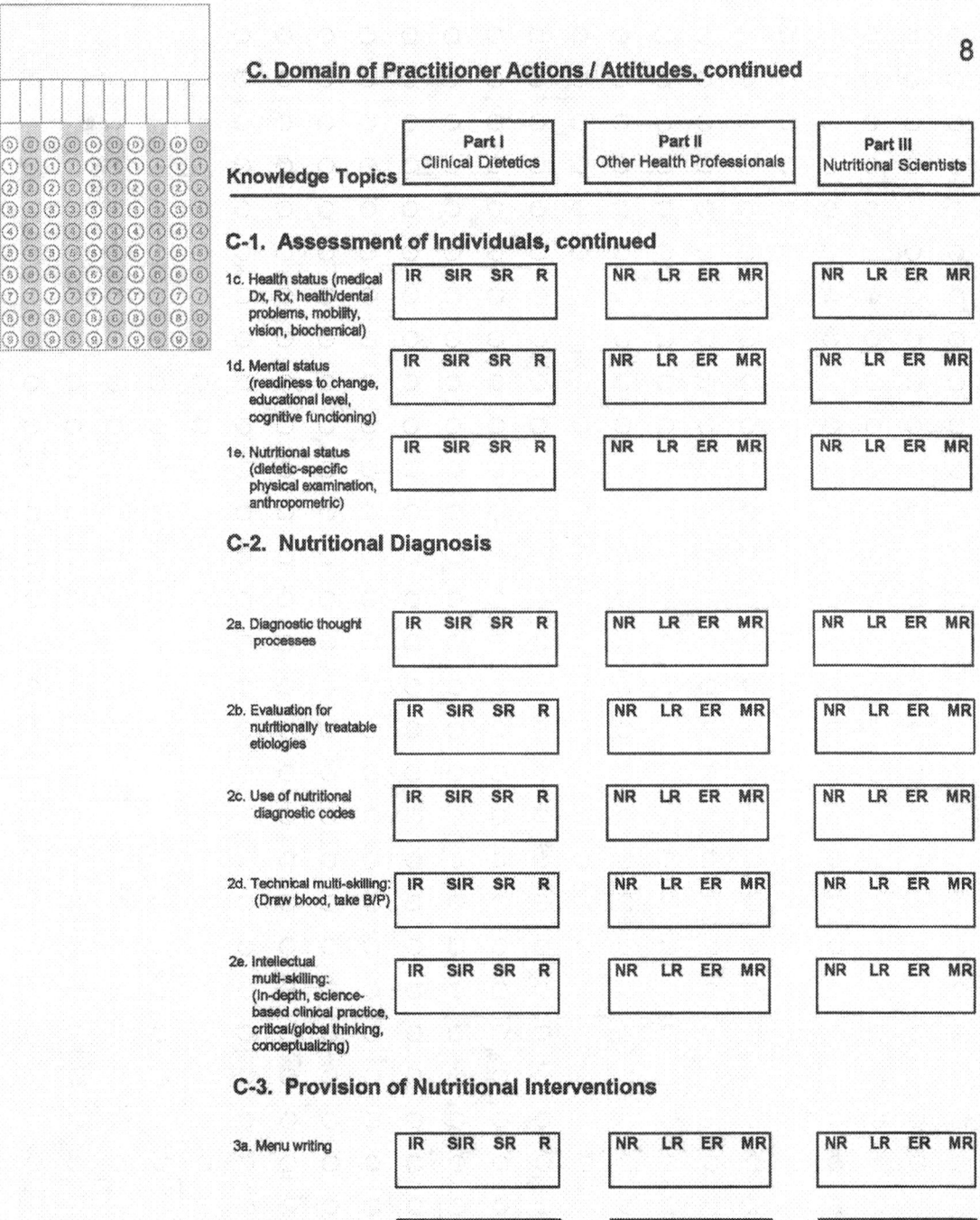

Knowledge Topics	Part I Clinical Dietetics	Part II Other Health Professionals	Part III Nutritional Scientists

C-1. Assessment of Individuals, continued

	Part I	Part II	Part III
1c. Health status (medical Dx, Rx, health/dental problems, mobility, vision, biochemical)	IR SIR SR R	NR LR ER MR	NR LR ER MR
1d. Mental status (readiness to change, educational level, cognitive functioning)	IR SIR SR R	NR LR ER MR	NR LR ER MR
1e. Nutritional status (dietetic-specific physical examination, anthropometric)	IR SIR SR R	NR LR ER MR	NR LR ER MR

C-2. Nutritional Diagnosis

	Part I	Part II	Part III
2a. Diagnostic thought processes	IR SIR SR R	NR LR ER MR	NR LR ER MR
2b. Evaluation for nutritionally treatable etiologies	IR SIR SR R	NR LR ER MR	NR LR ER MR
2c. Use of nutritional diagnostic codes	IR SIR SR R	NR LR ER MR	NR LR ER MR
2d. Technical multi-skilling: (Draw blood, take B/P)	IR SIR SR R	NR LR ER MR	NR LR ER MR
2e. Intellectual multi-skilling: (In-depth, science-based clinical practice, critical/global thinking, conceptualizing)	IR SIR SR R	NR LR ER MR	NR LR ER MR

C-3. Provision of Nutritional Interventions

	Part I	Part II	Part III
3a. Menu writing	IR SIR SR R	NR LR ER MR	NR LR ER MR
3b. Nutritional education	IR SIR SR R	NR LR ER MR	NR LR ER MR

C. Domain of Practitioner Actions/Attitudes, continued

Knowledge Topics	Part I Clinical Dietetics	Part II Other Health Professionals	Part III Nutritional Scientists

C-3. Provision of Nutritional Interventions, continued

	Part I	Part II	Part III
3c. Nutritional counseling	IR SIR SR R	NR LR ER MR	NR LR ER MR
3d. Nutritional goal-setting	IR SIR SR R	NR LR ER MR	NR LR ER MR
3e. Nutritional recommendations	IR SIR SR R	NR LR ER MR	NR LR ER MR
3f. Use of dietetic treatment protocols	IR SIR SR R	NR LR ER MR	NR LR ER MR

C-4. Documentation of Activities

	Part I	Part II	Part III
4a. Documentation of clinical dietetic activities	IR SIR SR R	NR LR ER MR	NR LR ER MR
4b. Measurement of dietetic outcomes	IR SIR SR R	NR LR ER MR	NR LR ER MR
4c. Management of clinical dietetic services	IR SIR SR R	NR LR ER MR	NR LR ER MR

C-5. Participation in Clinical Dietetic Research

	Part I	Part II	Part III
5a. As collaborator	IR SIR SR R	NR LR ER MR	NR LR ER MR
5b. As principal investigator	IR SIR SR R	NR LR ER MR	NR LR ER MR

C-6. Philosophy of Clinical Dietetics

	Part I	Part II	Part III
6a. Practitioner's personal professional philosophy	IR SIR SR R	NR LR ER MR	NR LR ER MR
6b. American Dietetic Association philosophy	IR SIR SR R	NR LR ER MR	NR LR ER MR

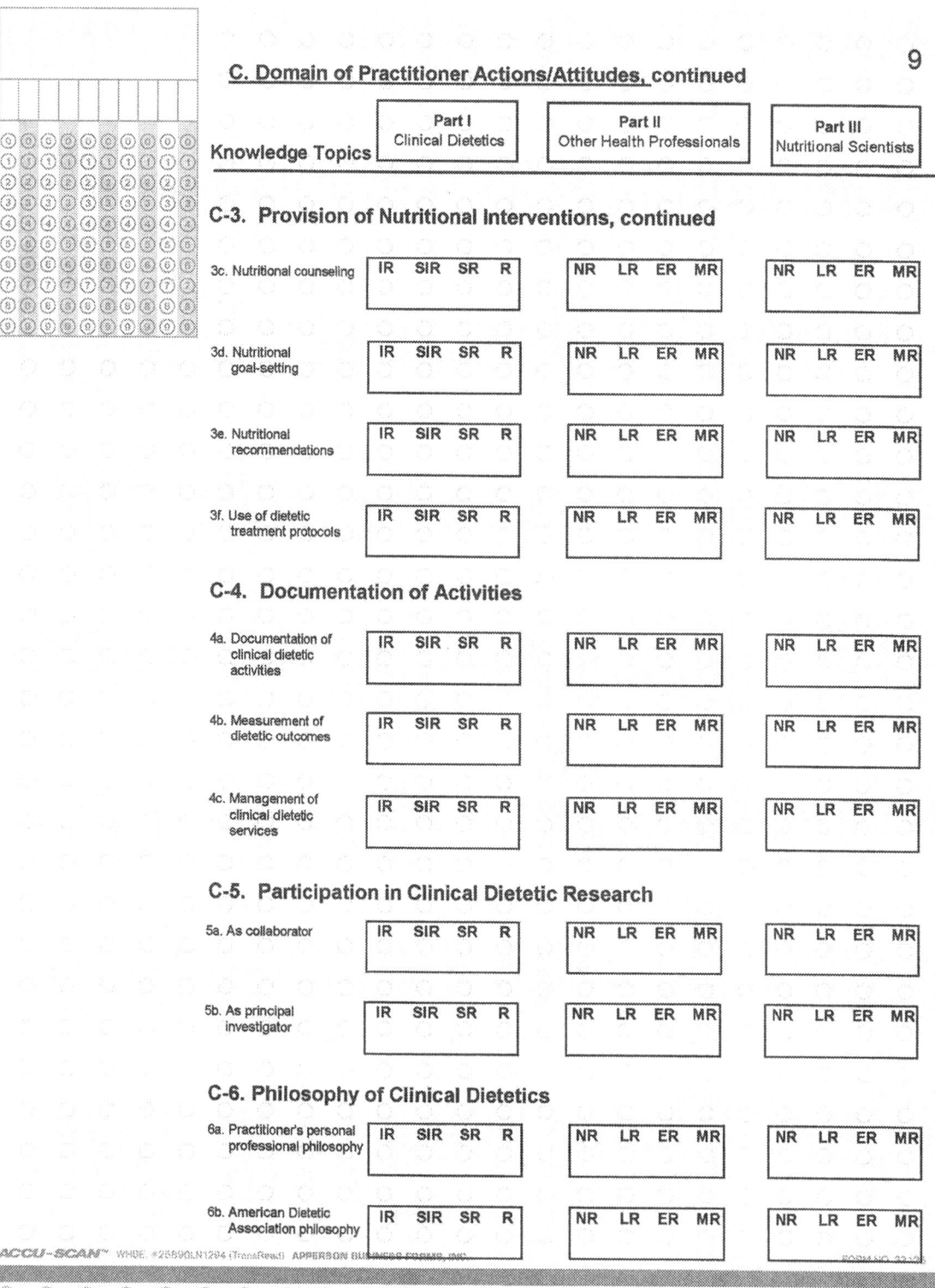

C. Domain of Practitioner Actions/Attitudes, continued

Knowledge Topics	Part I Clinical Dietetics	Part II Other Health Professionals	Part III Nutritional Scientists

C-7. Relationships

7a. Transformational leadership	IR SIR SR R	NR LR ER MR	NR LR ER MR
7b. Networks	IR SIR SR R	NR LR ER MR	NR LR ER MR
7c. Interdependent colleagues	IR SIR SR R	NR LR ER MR	NR LR ER MR
7d. Dietitian-Client relationships	IR SIR SR R	NR LR ER MR	NR LR ER MR

Which "other health professionals" were you considering when you completed this domain? Indicate any that apply.

Nurses ☐ Pharmacists ☐ Physicians ☐ Psychologists ☐ Other (please specify_____) ☐

D. Domain of Practitioner Environment

Practitioner Environment refers to the complex social and physical circumstances in which clinical dietetics is practiced. It includes relationships with other professionals, the local and national organization, the prevailing political and social milieu, the scientific knowledge, and the state of technology available and profession-specific tools.

D-1. Professional Credentialing

1a. Registered status -ADA	IR SIR SR R	NR LR ER MR	NR LR ER MR
1b. Board Certified Specialists - ADA	IR SIR SR R	NR LR ER MR	NR LR ER MR
1c. Fellow Status - ADA	IR SIR SR R	NR LR ER MR	NR LR ER MR
1d. RD Licensure -State	IR SIR SR R	NR LR ER MR	NR LR ER MR

D. Domain of Practitioner Environment, continued

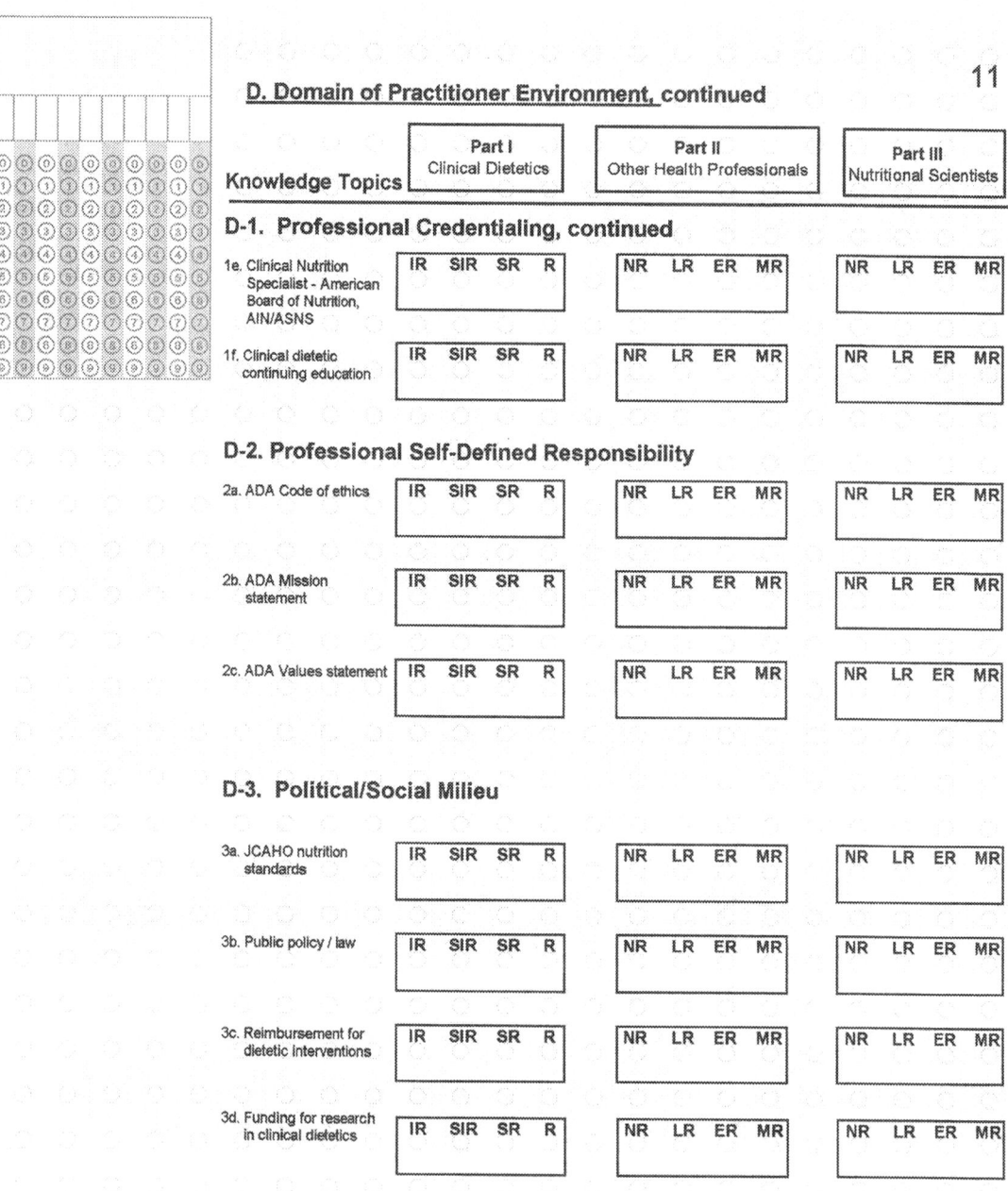

Knowledge Topics	Part I Clinical Dietetics	Part II Other Health Professionals	Part III Nutritional Scientists

D-1. Professional Credentialing, continued

	Part I	Part II	Part III
1e. Clinical Nutrition Specialist - American Board of Nutrition, AIN/ASNS	IR SIR SR R	NR LR ER MR	NR LR ER MR
1f. Clinical dietetic continuing education	IR SIR SR R	NR LR ER MR	NR LR ER MR

D-2. Professional Self-Defined Responsibility

	Part I	Part II	Part III
2a. ADA Code of ethics	IR SIR SR R	NR LR ER MR	NR LR ER MR
2b. ADA Mission statement	IR SIR SR R	NR LR ER MR	NR LR ER MR
2c. ADA Values statement	IR SIR SR R	NR LR ER MR	NR LR ER MR

D-3. Political/Social Milieu

	Part I	Part II	Part III
3a. JCAHO nutrition standards	IR SIR SR R	NR LR ER MR	NR LR ER MR
3b. Public policy / law	IR SIR SR R	NR LR ER MR	NR LR ER MR
3c. Reimbursement for dietetic interventions	IR SIR SR R	NR LR ER MR	NR LR ER MR
3d. Funding for research in clinical dietetics	IR SIR SR R	NR LR ER MR	NR LR ER MR
3e. Links with business / industry	IR SIR SR R	NR LR ER MR	NR LR ER MR

D. Domain of Practitioner Environment ,continued

Knowledge Topics	Part I Clinical Dietetics	Part II Other Health Professionals	Part III Nutritional Scientists

D-4. Available Tools and Technology

	Part I	Part II	Part III
4a. Tools for physical nutritional assessment of individuals	IR SIR SR R	NR LR ER MR	NR LR ER MR
4b. Tools for assessment of nutrition-related behaviors of individuals	IR SIR SR R	NR LR ER MR	NR LR ER MR
4c. Tools for attitude assessment of individuals	IR SIR SR R	NR LR ER MR	NR LR ER MR
4d. Computer programs re: nutrition and/or diet	IR SIR SR R	NR LR ER MR	NR LR ER MR
4e. Electronic communication capability	IR SIR SR R	NR LR ER MR	NR LR ER MR
4e. Nutrition education materials	IR SIR SR R	NR LR ER MR	NR LR ER MR

Which "other health professionals" were you considering when you completed this domain? Indicate any that apply.

Nurses Pharmacists Physicians Psychologists Other (please specify_____)

E. Domain of Client Actions / Attitudes

Client refers to any individual or group that present to a Clinical Dietitian for nutritional services.

Actions refer to behaviors engaged in or purposefully refrained from, relative to maintaining desired nutritional status or preventing nutrition-related departures from health, including but not limited to eating habits; food selection, procurement, intake; use of nutraceuticals, exercise, and utilization of recommendations.

Attitude refers to intrapersonal characteristics and processes of a client. Attitudes may be explicit and willfully affect behavior or implicit (unstated, unknown, subconscious or unconscious) and involuntarily affect behavior, not being under the influence of the will. In this instrument attitude includes, but is not limited to, ability, aptitude, beliefs, desires, emotions, ethics, ideas, knowledge, morals, opinions, preferences, thoughts, values, will and world view.

E-1. Acceptance of Responsibility for Self Care

	Part I	Part II	Part III
1a. Ability for nutritional self care	IR SIR SR R	NR LR ER MR	NR LR ER MR

E. Domain of Client Actions / Attitudes, continued

Knowledge Topics	Part I Clinical Dietetics	Part II Other Health Professionals	Part III Nutritional Scientists

E-1. Acceptance of Responsibility for Self Care

| 1b. Knowledge for nutritional self care | IR SIR SR R | NR LR ER MR | NR LR ER MR |
| 1c. World view re: health and nutrition | IR SIR SR R | NR LR ER MR | NR LR ER MR |

E-2. Influences on Food Intake

| 2a. Food preferences | IR SIR SR R | NR LR ER MR | NR LR ER MR |
| 2b. Emotional status | IR SIR SR R | NR LR ER MR | NR LR ER MR |

E-3. Choices Made

3a. Lifestyle choices

| | IR SIR SR R | NR LR ER MR | NR LR ER MR |

3b. Food choices

| | IR SIR SR R | NR LR ER MR | NR LR ER MR |

Which "other health professionals" were you considering when you completed this domain? Indicate any that apply.

Nurses   Pharmacists   Physicians   Psychologists   Other (please specify_____)

□   □   □   □   □

F. Domain of Client Environment

Client environment refers to the complex social and physical circumstances surrounding a client who receives nutritional interventions from a clinical dietitian. It includes influences of health status and medications, family and associates, food available, work, finances, cultural influences, the marketplace and self care skills

F-1. Food Resources

| 1a. Food sources | IR SIR SR R | NR LR ER MR | NR LR ER MR |
| 1b. Route of food delivery (enteral, parenteral) | IR SIR SR R | NR LR ER MR | NR LR ER MR |

F. Domain of Client Environment, continued

Knowledge Topics	Part I Clinical Dietetics	Part II Other Health Professionals	Part III Nutritional Scientists

F-2. Personal Social Resources

	Part I	Part II	Part III
2a. Adequate finances for food	IR SIR SR R	NR LR ER MR	NR LR ER MR
2b. Transportation to acquire food	IR SIR SR R	NR LR ER MR	NR LR ER MR
2c. Social milieu for eating	IR SIR SR R	NR LR ER MR	NR LR ER MR
2d. Care-giver who is responsible and knowledgeable re: nutrition needs	IR SIR SR R	NR LR ER MR	NR LR ER MR

F-3. Environmental Threats

	Part I	Part II	Part III
3a. Food-borne environmental toxins	IR SIR SR R	NR LR ER MR	NR LR ER MR
3b. Misleading nutritional claims	IR SIR SR R	NR LR ER MR	NR LR ER MR

Which "other health professionals" were you considering when you completed this domain? Indicate any that apply.

Nurses ☐   Pharmacists ☐   Physicians ☐   Psychologists ☐   Other (please specify_____) ☐

G. Domain of Nutraceuticals

Nutraceuticals refers to any substance that can be considered to be a food or a component of a food that affects health, including the prevention and treatment of disease. Such products range from all natural, processed, created / engineered / manufactured foods, designer food, functional foods, phytochemicals, isolated nutrients / supplements, chemopreventive agents and pharmafoods.

G-1. Knowledge Concerning Nutraceuticals

	Part I	Part II	Part III
1a. Composition of nutraceuticals	IR SIR SR R	NR LR ER MR	NR LR ER MR
1b. Processing of nutraceuticals	IR SIR SR R	NR LR ER MR	NR LR ER MR

G. Domain of Nutraceuticals, continued

Knowledge Topics	Part I Clinical Dietetics	Part II Other Health Professionals	Part III Nutritional Scientists

G-2. Appropriate Use of Nutraceuticals

	Part I	Part II	Part III
2a. Functions of nutraceuticals	IR SIR SR R	NR LR ER MR	NR LR ER MR
2b. Effects of nutraceuticals	IR SIR SR R	NR LR ER MR	NR LR ER MR

Which "other health professionals" were you considering when you completed this domain? Indicate any that apply.

Nurses ☐   Pharmacists ☐   Physicians ☐   Psychologists ☐   Other (please specify_____) ☐

Do you perceive the domains in the proposed Metaparadigm of Clinical Dietetics as encompassing the body of knowledge utilized in Clinical Dietetics?   YES   NO

If your answer was No, what do you perceive as being omitted?

Domains omitted:
_____
_____

Knowledge Topics omitted:
_____
_____

In which domain(s) of the proposed Metaparadigm of Clinical Dietetics would you place your work? (Indicate as many choices as you think applicable to your work) To review Domain definitions, see survey pages noted below:

Reference Person (Page 4)	Human Condition (Page 6)	Practitioner Actions / Attitudes (Page 7)	Practitioner Environment (Page 10)	Client Actions / Attitudes (Page 12)	Client Environment (Page 13)	Nutraceuticals (Page 14)
☐	☐	☐	☐	☐	☐	☐

Information About Your Practice of Clinical Dietetics

1. How long have you been practicing clinical dietetics?

<1 year	1 - 5 years	6 - 10 years	11 - 20 years	>20 years
☐	☐	☐	☐	☐

2. What is your primary (>50%) practice environment?

Hospital	HMO	Private Practice	WIC
☐	☐	☐	☐

Other _____

3. What is your highest degree?

BA	BS	MA	MS	MPH	PhD	DSc	DEd

Other _____

4. What was your route of entry into clinical dietetics?

Internship	CUP	Experience	AP4
☐	☐	☐	☐

Other _____

5. Have you published any papers, books, etc., while practicing clinical dietetics?

YES	NO

If Yes, how many?

1 - 2	3 - 6	7 - 10	>10

6. Have you been involved in clinical dietetic research?

YES	NO

If Yes:

PAST	PRESENT

7. Age:

20 - 29	30 - 39	40 - 49	50 - 59	60 - 69	70 - 79	> 79 years
☐	☐	☐	☐	☐	☐	☐

8. Gender:

Female	Male
☐	☐

Thank you for taking the time to complete this survey.

Requests for results: I would like to receive a summary of the survey results when the data analysis is completed:

YES	NO

Please mail the completed survey in the postage-paid envelope to

Ruth Leyse, MS, RD
309 Shantz Hall
Department of Nutritional Sciences
University of Arizona
Tucson, AZ 85721-0038

978-0-595-42205-0
0-595-42205-5

www.ingramcontent.com/pod-product-compliance
Lightning Source LLC
Chambersburg PA
CBHW081146180526
45170CB00006B/1944